T0135029

Mathematics and Its Applications (*Soviet Series*)

Bayesian Approach to Global Optimization

Theory and Applications

by

Jonas Mockus

Academy of Sciences of the Lithuanian SSR,
Institute of Mathematics and Cybernetics, Vilnius, U.S.S.R.

KLUWER ACADEMIC PUBLISHERS
DORDRECHT / BOSTON / LONDON

Library of Congress Cataloging in Publication Data

Mockus, Jonas.
    Bayesian approach to global optimization.

    (Mathematics and its applications. Soviet series)
    Includes index.
    1. Mathematical optimisation. 2. Bayesian
statistical decision theory. I. Title. II. Series:
Mathematics and its applications (D. Reidel Publishing
Company). Soviet series.
QA402.5.M58   1989   519   88-27188

ISBN-13: 978-94-010-6898-7   e-ISBN-13: 978-94-009-0909-0
DOI: 10.1007/978-94-009-0909-0

Published by Kluwer Academic Publishers,
P.O. Box 17, 3300 AA Dordrecht, The Netherlands.

Kluwer Academic Publishers incorporates
the publishing programmes of
D. Reidel, Martinus Nijhoff, Dr W. Junk and MTP Press.

Sold and distributed in the U.S.A. and Canada
by Kluwer Academic Publishers,
101 Philip Drive, Norwell, MA 02061, U.S.A.

In all other countries, sold and distributed
by Kluwer Academic Publishers Group,
P.O. Box 322, 3300 AH Dordrecht, The Netherlands.

SERIES EDITOR'S PREFACE

Mathematics is a tool for thought. A highly necessary tool in a world where both feedback and non-linearities abound. Similarly, all kinds of parts of mathematics serve as tools for other parts and for other sciences.

Applying a simple rewriting rule to the quote on the right above one finds such statements as: 'One service topology has rendered mathematical physics ...'; 'One service logic has rendered computer science ...'; 'One service category theory has rendered mathematics ...'. All arguably true. And all statements obtainable this way form part of the raison d'être of this series.

This series, *Mathematics and Its Applications*, started in 1977. Now that over one hundred volumes have appeared it seems opportune to reexamine its scope. At the time I wrote

> "Growing specialization and diversification have brought a host of monographs and textbooks on increasingly specialized topics. However, the 'tree' of knowledge of mathematics and related fields does not grow only by putting forth new branches. It also happens, quite often in fact, that branches which were thought to be completely disparate are suddenly seen to be related. Further, the kind and level of sophistication of mathematics applied in various sciences has changed drastically in recent years: measure theory is used (non-trivially) in regional and theoretical economics; algebraic geometry interacts with physics; the Minkowsky lemma, coding theory and the structure of water meet one another in packing and covering theory; quantum fields, crystal defects and mathematical programming profit from homotopy theory; Lie algebras are relevant to filtering; and prediction and electrical engineering can use Stein spaces. And in addition to this there are such new emerging subdisciplines as 'experimental mathematics', 'CFD', 'completely integrable systems', 'chaos, synergetics and large-scale order', which are almost impossible to fit into the existing classification schemes. They draw upon widely different sections of mathematics."

By and large, all this still applies today. It is still true that at first sight mathematics seems rather fragmented and that to find, see, and exploit the deeper underlying interrelations more effort is needed and so are books that can help mathematicians and scientists do so. Accordingly MIA will continue to try to make such books available.

If anything, the description I gave in 1977 is now an understatement. To the examples of interaction areas one should add string theory where Riemann surfaces, algebraic geometry, modular functions, knots, quantum field theory, Kac-Moody algebras, monstrous moonshine (and more) all come together. And to the examples of things which can be usefully applied let me add the topic 'finite geometry'; a combination of words which sounds like it might not even exist, let alone be applicable. And yet it is being applied: to statistics via designs, to radar/sonar detection arrays (via finite projective planes), and to bus connections of VLSI chips (via difference sets). There seems to be no part of (so-called pure) mathematics that is not in immediate danger of being applied. And, accordingly, the applied mathematician needs to be aware of much more. Besides analysis and numerics, the traditional workhorses, he may need all kinds of combinatorics, algebra, probability, and so on.

In addition, the applied scientist needs to cope increasingly with the nonlinear world and the extra mathematical sophistication that this requires. For that is where the rewards are. Linear models are honest and a bit sad and depressing: proportional efforts and results. It is in the non-linear world that infinitesimal inputs may result in macroscopic outputs (or vice versa). To appreciate what I am hinting at: if electronics were linear we would have no fun with transistors and computers; we would have no TV; in fact you would not be reading these lines.

There is also no safety in ignoring such outlandish things as nonstandard analysis, superspace and anticommuting integration, p-adic and ultrametric space. All three have applications in both electrical engineering and physics. Once, complex numbers were equally outlandish, but they frequently proved the shortest path between 'real' results. Similarly, the first two topics named have already provided a number of 'wormhole' paths. There is no telling where all this is leading - fortunately.

Thus the original scope of the series, which for various (sound) reasons now comprises five subseries: white (Japan), yellow (China), red (USSR), blue (Eastern Europe), and green (everything else), still applies. It has been enlarged a bit to include books treating of the tools from one subdiscipline which are used in others. Thus the series still aims at books dealing with:

- a central concept which plays an important role in several different mathematical and/or scientific specialization areas;
- new applications of the results and ideas from one area of scientific endeavour into another;
- influences which the results, problems and concepts of one field of enquiry have, and have had, on the development of another.

Trying to optimize something - if possible everything - is one of the oldest preoccupations of human kind. It is also one of the oldest parts of mathematics to try to find all kinds of extrema. And certainly from the applied point of view it is easy to see that this part of mathematics at least ought to be applicable to virtually everything. And so it is, though it required widespread use of computing power to really flourish. Relatively easily formulated examples of optimization problems are assignment problems, optimal cooling of steel labs in a rolling mill, optimal design of a road network, allocation and location problems in economics, and other combinatorial optimization problems. Less obvious but equally relevant examples are problems in VLSI chip design and databases which can be formulated as (constrained) global optimization problems (with a quadratic objective function).

Local optimization is better developed (mathematically) but it is not good enough for these problems and global optimization requires additional ideas which go beyond calculus and variational principles. Computing power alone is not nearly enough. Even very simple optimization problems have a habit of becoming rapidly too large to be handled by routine programming. And so, thanks to computing power, rather than the reverse, a large number of new areas in pure and applied mathematics came into being. Global optimization, the topic of this book, is one of them. More particularly this book is about the Bayesian approach to this large and varied problem; one of the more successful and promising approaches. Together with the theory there are a large number of technical and industrial applications and there is also a floppy disc (MS-DOS format; FORTRAN programs) with all the programs so that the reader can test and try out these ideas and programs on his own optimization problems. No doubt many will do so to experience for themselves the power (and elegance) of these methods.

Perusing the present volume is not guaranteed to turn you into an instant expert, but it will help, though perhaps only in the sense of the last quote on the right below.

The shortest path between two truths in the real domain passes through the complex domain.

J. Hadamard

La physique ne nous donne pas seulement l'occasion de résoudre des problèmes ... elle nous fait pressentir la solution.

H. Poincaré

Never lend books, for no one ever returns them; the only books I have in my library are books that other folk have lent me.

Anatole France

The function of an expert is not to be more right than other people, but to be wrong for more sophisticated reasons.

David Butler

Bussum, October 1988

Michiel Hazewinkel

CONTENTS

Preface

Global optimization is a subject of tremendous potential application. The past two decades have witnessed increasing efforts in research and application. The result is a collection of methods and algorithms. Progress in the computational aspects of global optimization has been achieved, but it is not however, as impressive as could have been expected considering the development in power of digital computers and the vast field of possible applications.

The possible explanation is that in global optimization there exists a wide gap between theory and applications, between mathematical and heurristic methods.

This book is intended to narrow the gap by means of a new approach to the development of numerical methods of global optimization. The idea of this approach is to develop methods of optimization which are the best in the senes of average deviation from the global minimum. To define the average deviation some *a priori* distribution should be fixed. This is a distinctive property of the Bayesian approach.

A balanced coverage of the theory, applications and computations of global optimization are given.

The general orientation of the presentation is to demonstrate the potential computational advantages of the Bayesian approach in a manner which may assist the practitioner in solving problems in real life applications. The corresponding portable software is included in addition to the theory, the description of methods, algorithms and a set of practical examples.

The advancement of global optimization promises to expand the use of optimization methods into new applications, the important examples being the CAD/CAM systems ad the optimal design of experiments.

Chapter 1 is a brief outline of the main advantages of the Bayesian approach.

Chapter 2 presents a general definition of Bayesian methods of global optimization.

Chapter 3 gives the axiomatic non-probabilistic justification of the Bayesian approach by the system of clear and simple assumptions concerning the subjective preferences.

In Chapter 4 the family of *a priori* stochastic models which provide the convergence of Bayesian mehods to the global minimum of any continuous function is considered. A Gaussian stochastic model is derived from the conditions of homogeneity, independence of partial differences and continuity of sample functions.

Chapter 5 provides the expressions for the one-step approximation of the dynamic programming equations corresponding to the Bayesian methods in the Gaussian case. This chapter describe one of the most unconventional results of the book: the new nonclassical stochasitc model, where the usual consistency conditions of Kolmogorov are replaced by the weaker condition of the risk function continuity. It gives the posssibility of avoiding the computational difficutlties connected with the inversion of matrices of very high order, which is necessary using the classical statistical models. The results of the international 'competition' of algorithms of global optimization are discussed.

Chapter 6 discusses the methods and algorithms of reducing the dimensionality of the problems of global optimization in a way that minimizes the average deviation from the true solution.

Chapter 7 provides an example of how the Bayesian approach can be used to increase the efficiency of the methods of stohastic apporximation to find local minima of functions with 'noise'.

Chapter 8 describes th application of Bayesian methods in real-life applicationa of engineering design and experimental planning.

Chapter 9 provides the description of the Portable FORTRAN package for global optimization, the complete listing of which is given in the Appendix.

In the book only the case of continuous variables is considered becauses, in this case, the neighbourhood is uniquely defined. The methods developed for continuous variables can also be applied to the discrete case if the rounding error can be regarded as negligible.

I am gratefully indebted to my colleagues of the Optimal Decision Theory Department of the Institute of Mathematics and Cybernectics of the Academy of Sciences of the Lithuanian S.S.R. who contributed to the development and application of the methods of global optimization.

CHAPTER 1

GLOBAL OPTIMIZATION AND THE BAYESIAN APPROACH

1.1 What is global optimization?

Any decision problem (with an objective function f to be minimized or maximized) may be classified as a global optimization problem, if there is no additional information indicating that there is only one minimum (or maximum). This definition also includes the case of discrete optimization, when the (quantifiable) decision variables must assume discrete values. In this book we shall consider only the case of continuous variables because for the purpose of developing solutions the discrete optimization problems are better regarded separately.

Global optimization as defined above is not a new mathematical subject. In the classical textbooks the global optimization problem is usually dismissed by statements such as: "If there are many minima (or maxima) they should all be compared". The statement is obviously correct if it is a practical possibility

1) to find any finite number of local minima

2) to prove that no more local minima exist.

Such possibilities exist only in some special cases.

Generally we should abandon the hope of finding the exact global minimum and seek some approximations to it. In such a case the main theoretical and practical problem is how to define the approximate solutions and how to find them.

1.2 Advantages of the Bayesian approach to global optimzation

Since only an approximation to the global minimum can usually be defined we shall classify different methods depending on the definition of the deviation from the exact solution. In this way two main approaches can be formulated. The first one can be called the worst case analysis, when the deviation for the worst possible condition is

$$\sup_{f \in F} |f(x) - f(x_0)| \qquad (1.2.1)$$

1

Here $f = f(x) \in F$ is the function to be minimized,

    $x \in A \subset R^m$ is the vector of variables,

    A is the feasible set,

    $|f(x) - f(x_0)|$ is the 'distance' between x and x_0,

    x_0 is the global minimum,

    F is the family of functions f.

The more usual definition of the distance between the points x and x_0, namely $\|x - x_0\|$ leads to confusion if there is more than one global minimum because then the distance $\|x - x_0\|$ is not uniquely defined.

The advantage of definition (1.2.1) is that it gives the exact upper bound of the error of the method, but it has a disadvantage too, because the expression (1.2.1) can go to infinity except in special cases such as the Lifshitzian functions $f(x)$. In this sense it is more convenient to use the second approach and to define the deviation as an average 'distance' from the global minimum, namely

$$\int_F \left(f(x) - f(x_0) \right) P(df) \tag{1.2.2}$$

Here P is some σ-additive measure which should be fixed *a priori*. It is convenient to regard P as a probability measure defined on a family of functions to be minimized by a given method. The measure P has a similar meaning to the *a priori* probability in Bayesian decision theory so the second approach is called the Bayesian approach.

The Bayesian definition of deviation (1.2.2) explains the reason for the wide application of heuristic methods in global optimization. It almost corresponds to the usual definition of heuristic methods, as such methods, which work reasonably well in most real-life applications can, nevertheless, be very bad in some exceptional cases.

The definition (1.2.2) can serve as a reasonable estimation of the efficiency of the heuristic procedures and so help to narrow the gap between the heuristic and the mathematical methods and to develop the efficient methods of global optimization which are based on clear mathematical assumptions.

The important property of the Bayesian approach is that the Bayesian method which minimizes the deviation (1.2.2) depends on the *a priori* distribution P. It is both the main advantage and the main disadvantage of the Bayesian approach. The advantage is that we can develop methods in accordance with average properties of the function to be minimized. The disadvantage is the arbitrariness and uncertainty of how to fix the *a priori* distribution P, especially when the function $f(x)$ describes some unique object and so the probability P cannot be derived from the corresponding frequencies.

To decrease the level of uncertainty some conditions are defined such as the convergence of Bayesian methods to a global minimum of any continuous function. The convergence conditions narrow the family of feasible *a priori* distributions. It can be narrowed even further if some additional conditions are introduced. For example, from the conditions of continuity of sample functions, homogeneity of P and independence of partial differences, it conforms to the Gaussian P with constant mean μ and variance σ^2 and the covariance between the points x_j and x_k.

$$\sigma_{jk} = \sigma^2 \prod_{i=1}^{m} \left(1 - \frac{|x_j^i - x_k^i|}{2} \right), \quad -1 \leq x^i \leq 1, \quad i = 1, \dots, m$$

To justify the application of the probabilities in the cases when we are dealing, not actually with probability but with uncertainty, the system of clear conditions about the subjective preferences is given, when there exists a unique *a priori* density function corresponding to given subjective preferences.

Any result which narrows the gap between the theory and practice of global optimization is considered as most important because it is hardly possible to prove the practical efficiency of methods and algorithms following only theoretical considerations. It does not seem possible to do it empirically either. So the way remaining is to use both theoretical and empirical means considering not only the well-known text book test functions but also real-life examples.

To make theoretical methods computationally feasible, the theoretical models are formulated in terms of simple and clear basic assumptions. Those assumptions which make the models more complicated and are not absolutely necessary, are replaced by more convenient conditions, regardless of their historic importance or logical elegance. The important example in this book is the replacement of the consistency condition of Kolmogorov by the condition of the continuity of the Bayesian risk function.

CHAPTER 2

THE CONDITIONS OF BAYESIAN OPTIMALITY

2.1. Introduction

Denote the objective function which we shall minimize as

$$f = f(x) = f(x, \omega), \quad x \in A., \quad \omega \in \Omega \tag{2.1.1}$$

Suppose that f is a continuous function of x and measurable function of ω,

$$A \subset R^m. \tag{2.1.2}$$

Here A is a compact set and Ω is a set of indices ω corresponding to all continuous functions of $x \in A$.

Assume that we can observe (calculate or define by physical experimentiation) the values of $f(x_i)$ at the points x_i. The results of the observations will be denoted as

$$y_i = f(x_i), \quad i = 1, \dots, N \tag{2.1.3}$$

where N is the total number of observations.

Denote the vector of observations as

$$z_n = (x_i, y_i, \ i = 1, \dots, n). \tag{2.1.4}$$

Define the decision function as

$$d_n = d_n(z_n), n = 0, \dots, N \tag{2.1.5}$$

where d_n is a mapping of $(A \times R)^n$ into A.

Denote the sequence of decision function as

$$d = (d_0, \dots, d_N). \tag{2.1.6}$$

Suppose that

$$d_n \in D_n, \ n = 0, \ldots, N$$

where D_n is the set of all measurable mappings of $(A \times R)^n$ into A and

$$d \in D = \underset{n=0}{\overset{N}{\times}} D_n.$$

Assume that the observation points are defined by the decision function d:

$$x_{i+1} = x_{i+1}(d) = d_i(z_i), \ i = 0, \ldots N. \tag{2.1.7}$$

Denote as δ the loss function which is the deviation from the minimum of $f(x)$ when the sequence of decision functions d is used:

$$\delta = \delta(d) = \delta(d, \omega) = f(x_{N+1}, \omega) - \underset{x \in A}{\min} f(x, \omega). \tag{2.1.8}$$

Here x_{N+1} is the point of the final decision which should be made after all N observations are completed.

The average deviation $R_0(d)$ (the risk) can be expressed as the Lebesgue integral:

$$R_0(d) = E\{\delta(d)\} = \int_\Omega \delta(d\omega) \, P(d\omega). \tag{2.1.9}$$

Here P is a probability measure defined on Borel sets of Ω.

Taking into account (2.1.8)

$$R_0(d) = \int_\Omega f(x_{N+1}(d), \omega) \, P(d\omega) - \int_\Omega \underset{x \in A}{\min} f(x, \omega) \, P(d\,\omega). \tag{2.1.10}$$

The decision function $d' \in D$ was called, by Mockus (1972), the Bayesian method if

$$R(d') = \int_\Omega f(x_{N+1}(d'), \omega) \, P(d\omega) = \underset{d \in D}{\inf} \int_\Omega f(x_{N+1}(d), \omega) \, P(d\omega). \tag{2.1.11}$$

The condition (2.1.11) minimizes the expected deviation (2.1.10) if

$$\left| \int_\Omega \underset{x \in A}{\min} f(x, \omega) \, P(d\,\omega) \right| < \infty \tag{2.1.12}$$

as the second integral in (2.1.10) does not depend on d.

The definition (2.1.11) is more general than that based on the minimization of the expected deviation (2.1.10) because the condition (2.1.11) can also be applied to cases when (2.1.12) is violated since the expected minimum is not bounded.

2.2 Reduction to dynamic programming equations

It is convenient to reduce, see Mockus (1969, 1972) the condition (2.1.11) to recurrent equations of dynamic programming.

Suppose there exists a decision function $d'' \in D$ satisfying the recurrent equations

$$u_N(z_N) = \inf_{x \in A} E\{f(x)/z_N\}$$

$$u_{n-1}(z_{n-1}) = \inf_{x \in A} E\{u_n(z_{n-1}, x, f(x)/z_{n-1})\}, \quad n = N, \dots, 2$$

$$u_0 = \inf_{x \in A} E\{u_1(x, f(x))\} \tag{2.2.1}$$

where $E\{f(x)/z_N\}$ denotes the conditional expectation of the random variable $f(x)$ with regard to the random vector z_N and the n-th equation defines the relation between x_{n+1} and z_n, thus defining the n-th component of d''_n of the decision function d''. The relation between the Bayesian solution d' and d'' is shown by the following theorem of Mockus (1972).

THEOREM 2.2.1. *If*

$$E\{|f(x_{N+1}(d))|\} < \infty \tag{2.2.2}$$

for any $d \in D$ and there exists the solution d'' of (2.2.1) and

$$d'' \in D$$

then d'' is the Bayesian method in the sense of (2.1.11).

Proof. Denote

$$u_N(z_N) = E\{f(d_N(z_N))/z_N\} \tag{2.2.3}$$

$$u_{n-1}^d(z_{n-1}) = E\{u_n(z_{n-1}, d_n(z_n), f(d_n(z_n))/z_{n-1})\}, \quad n = N, \ldots, 2$$

$$u_0^d = E\{u_1^d(d_1, f(d_1))\}. \tag{2.2.4}$$

By the recurrent substitution

$$u_0^d = E\{E \ldots E\{E\{f(d_N(z_N))/z_N\}/z_{N-1}\} \ldots /z_1\}. \tag{2.2.5}$$

It is well known that under condition (2.2.2) the expression (2.2.5) can be reduced to

$$u_0^d = E\{f(d_N(z_N))\} \tag{2.2.6}$$

or using the notation (2.1.7)

$$u_0^d = E\{f(x_{N+1}(d))\}. \tag{2.2.7}$$

As the solution of (2.2.4), we shall define a sequence of measurable decision functions $d = (d_0, \ldots, d_N)$ which minimize the conditional expectations with probability 1

$$u_N^d(z_N) = \inf_{x \in A} E\{f(x)/z_N\}$$

$$u_{n-1}^d(z_{n-1}) = \inf_{x \in A} E\{u_n^d(z_{n-1}, x, f(x)/z_{n-1})\}, \quad n = N, \ldots, 2$$

$$u_0^d = \inf_{x \in A} E\{u_1(x, f(x))\}. \tag{2.2.8}$$

In accordance with the conditions of Theorem 2.2.1 there exist such measurable decision functions

$$x_{N+1} = d_N(z_N)$$

$$x_n = d_{n-1}(z_{n-1}), \quad n = N, \ldots, 2$$

$$x_1 = d_0 \tag{2.2.9}$$

which minimize the conditional expectations (with regard to $x \in A$):

$$E\{f(x)/z\}$$

$$E\{u_n^d(z_{n-1}, x, f(x)/z_{n-1})\}, \quad n = N, \ldots, 2$$

$$E\{u_1^d(x, f(x))\}. \tag{2.2.10}$$

Comparing (2.2.4) and (2.2.8) we can write

$$u_N^{d''}(z_N) \le u_N^d(z_N)$$

$$u_{n-1}^{d''}(z_{n-1}) \le u_{n-1}^d(z_{n-1}), \quad n = N, \ldots, 2$$

$$u_0^{d''} \le u_0^d. \tag{2.2.11}$$

This means that

$$u_0^{d''} = \inf_{d \in D} u_0^d. \tag{2.2.12}$$

From (2.2.7)

$$\inf_{d \in D} u_0^d = \inf_{d \in D} \{f(x_{N+1}(d))\}. \tag{2.2.13}$$

From (2.2.13) and (2.1.11)

$$u_0^d = E\{f(x_{N+1}(d'))\}.$$

Minimization with probability 1 means that the equalities (2.2.8) can be violated on some subsets of values of vectors z_n and the *a priori* probability of such subsets is zero. We shall suppose that all *a priori* probability distributions are absolutely continuous and that the conditional density is uniquely defined by the usual ratio of the corresponding multidimensional densities. In such a case the condition 'with probability 1' can be omitted.

EXAMPLE 2.2.1 Suppose that

$$f(x) = (x - \omega)^2, \quad A = [-1, 1], \quad \Omega = [-1, 1], \quad N = 1$$

and the *a priori* density function is

$$p(\omega) = 1/2, \quad \omega \in \Omega$$

then the decision function $d'' = (d_0'', \ldots, d_N'')$ is defined by the equations

$$u_1(z_1) = \inf_{x \in A} E\{f(x)/z_1\}$$

$$u_0 = \inf_{x \in A} E\{u(x, f(x))\}.$$

Here the conditional expectation

$$E\{f(x)/z_1\} = \begin{cases} (x - \omega_1)^2, & \omega_1 \in \Omega, \omega_2 \bar{\in} \Omega \\ (x - \omega_2)^2, & \omega_1 \bar{\in} \Omega, \omega_2 \in \Omega \\ 1/2(x - \omega_1)^2 + 1/2\,(x - \omega_2)^2, & \omega_1 \in \Omega, \omega_2 \in \Omega \end{cases}$$

and the decision function

$$d_1(z_1) = \begin{cases} x_1, & \omega_1 \in \Omega, \omega_2 \in \Omega \\ \omega_1, & \omega_1 \in \Omega, \omega_2 \bar{\in} \Omega \\ \omega_2, & \omega_1 \bar{\in} \Omega, \omega_2 \in \Omega \end{cases}$$

where $\omega_1 = x_1 - \sqrt{(f(x_1))}, \; \omega_2 = x_1 + \sqrt{(f(x_1))}$.
Hence

$$u_1(z_1) = \begin{cases} f(x_1), & \omega_1 \in \Omega, \omega_2 \in \Omega \\ 0, & \text{in other cases,} \end{cases}$$

so

$$E\{u_1(x, f(x))\} = \int_{\Omega_x} (x - \omega)^2 \, d\omega$$

where

$$\Omega_x = [\max(2x - 1, -1), \; \min(2x + 1, 1)]$$

therefore

$$d'' = \pm 1$$

and

$u_0 = 0.$

Thus the point of the first observation is

$x_1(d'') = \pm 1,$

and the final decision is

$$x_2(d\,'') = \begin{cases} 1 - \sqrt{f(1)}, & x_1(d'') = 1, \\ -1 + \sqrt{f(-1)}, & x_1(d'') = -1. \end{cases}$$

EXAMPLE 2.2.2. The statistical model is

$$f(x) = x^2 + \omega x, \quad A = R, \quad \Omega = R, \quad N = 1,$$

where the *a priori* density function of the stochastic variable ω is $n(0, 1)$.

Here $n(0, 1)$ is the Gaussian density function. The expectation is zero and the standard deviation is 1.

We may observe the stochastic function

$$\phi(x) = f(x) + \eta,$$

where η is an independent stochastic variable - the noise - with the probability density function $n(0, 1)$.

In this case the conditional expectation is

$$E\{f(x)/z_1\} = x^2 - \frac{xx_1}{x_1^2 + 1} (\phi(x_1) - x_1^2),$$

and the decision function is

$$d''(z_1) = 1/2 \, \frac{x_1}{x_1^2 + 1} (\phi(x_1) - x_1^2),$$

hence

$$u_1(z_1) = -1/4 \, \frac{x_1^2 (\phi(x_1) - x_1^2)^2}{(x_1^2 + 1)^2}$$

thus

$$E\{u_1(x, \phi(x))\} = -1/4 \; \frac{x^2}{x^2 + 1}$$

The Bayesian decision function d'' does not exist. There exists the ε-Bayes decision function d_0^{ε}. If $\varepsilon = 1/8$, then

$$|x_1(d^{\varepsilon})| \geq 1.$$

If $x_1(d^{\varepsilon}) = 1$, then the final ε-Bayes decision is

$$x_2(d^{\varepsilon}) = 1/4(1 - \phi(1)).$$

EXAMPLE 2.2.3. Consider the case when

$$f(x) = f_i(x), \; x \in A_i, \quad \bigcup_{i=1}^{n} A_i = A, \; A_j \cap A_j = \varnothing$$

where $f_i(x)$ are independent stochastic functions.

Suppose that there are N observations which should all be uniformly distributed in on one of the sets A_i. Assume that we know the distribution functions

$$F_i(y) = F\{f_i(\xi) < y\},$$

where the distribution of ξ is uniform in A_i.

Then from (2.1.11) the Bayesian decision

$$d' \in \arg\min_i E_i\{y_{0N}\}$$

corresponding average deviation

$$R(d') = \min_i E_i\{y_{0N}\} = \min_i \int_{-\infty}^{\infty} y_{0N} \, d\psi_i(y_{0N}).$$

Here $y_{0N} = \min_{1 \leq j \leq N} y_j$ and $\psi_i(y_{0N})$ is a distribution function of the minimum.

In a case of the Gaussian distribution function with parameters μ_i, σ_i the expectation of the best observed value in the set A_i

$$E_i\{y_{0N}\} = \mu_i - b\sigma_i + 0(1/\ln N)$$

where

$$b = \sqrt{(2 \ln N)} - \frac{\ln \ln N + \ln 4\pi - 2(1 - b_0)}{2\sqrt{(2 \ln N)}}$$

Here $b_0 = 0.5772 \ldots$ is the Euler constant.

Neglecting the remainder term $0(\ln \ln N / \sqrt{(\ln N)})$ we can write

$$E_i\{y_{0N}\} = \mu_i - \sqrt{(2 \ln N)}\sigma_i.$$

In a case of lognormal distribution, see Mockus (1967), the expectation

$$E_i\{y_{0N}\} = \varepsilon_i + \Gamma \left(\frac{\sigma_i}{\sqrt{(2 \ln N)}} + 1\right) \exp (\mu_i - b\sigma_i).$$

Here $\Gamma(\cdot)$ is a gamma function.

Neglecting the remainder terms

$$E_i\{y_{0N}\} = \varepsilon_i + \exp (\mu_i - \sqrt{(2 \ln N)}\sigma_i)$$

where ε_i is the minimum of $f_i(x)$ in A_i.

If $0 \leq \sigma_i \leq \sqrt{(2 \ln N)}$, then the error of the last formula will not exceed 14%.

The lognormality of the distribution function $F_i(y)$ was noticed by Shaltenis and Mockus (1963), when the problem of the optimization of multi-stage development of rural electrical networks was investigated. The same distribution was observed by Mockus (1964), when the problem of the optimization of electrical meters was considered. The conditions of asymptotic normality and lognormality of $F_i(y)$ were given by Mockus (1964). There it was assumed that $f_i(x)$ could be represented as a sum, or a product, of components which depends on different variables. Later the conditions were extended by Mockus (1965) to the case where each of n components $f_i(x)$ can depend on $r = o(n^{1/3})$ variables. The proofs were given by Mockus (1966). Example 2.2.2 was described by Mockus (1963). It was apparently the first time that the Bayesian approach and a stochastic model of function $f(x)$ were applied to the problem of global optimization. In the same paper a form of adaptive Bayesian approach was also used when the parameters of the *a priori* distribution $F_i(y)$ had to be estimated using some additional observations. The adaptive Bayesian approach will be discussed later in 2.6.

2.3 The existence of a measurable solution

The conditions when there exists a measurable solution of the recurrent equation (2.2.1) were given by Mockus (1978):

THEOREM 2.3.1. *If the conditional expectations in the recurrent equations* (2.2.1) *are the continuous functions of x and z_n and A is a compact set, then there exists a measurable solution $d'' \in D_n$ of* (2.2.1).

Proof. Denote

$$D(z) = \arg \min_{x \in A} \psi(x, z), \tag{2.3.1}$$

where

$$\psi(x, z) = E\{u_{n+1}(z, x, f(x)|z)\} \tag{2.3.2}$$

and

$$z \in (A \times R)^n. $$

Here $D(z)$ is the mapping of $(A \times R)^n$ into the set of 2^A of all closed subsets of $A \subset R^m$.

We shall now prove that under the conditions of Theorem 2.3.1 the function $D(z)$ is upper semi-continuous.

The multi-valued function $D(z)$ is called upper semi-continuous if from the conditions

$$\lim_{k \to \infty} z^k = z^0, \tag{2.3.3}$$

$$\lim_{k \to \infty} x_k = x_0, \tag{2.3.4}$$

and

$$\psi(x_k, z^k) \le \psi(x, z^k), \quad x \in A \tag{2.3.5}$$

it follows that

$$\psi(x_0, z^0) \le \psi(x, z^0), \quad x \in A. \tag{2.3.6}$$

Since ψ is assumed to be a continuous function then from conditions (2.3.3), (2.3.4)and (2.3.5) there follows the limit condition (2.3.6). Hence the function $D(z)$ is upper semi-continuous.

It is well known, see Kuratowski (1968), that if the multi-valued function $D(z)$ is upper semi-continuous then there exists a selector of class 1, namely a single-valued function, such that the values $d(z) \in D(z)$ and the originals of open sets are F_σ-sets (which consist of closed sets and their countable unions). Since F_σ-sets are Borel sets, see Kuratowski (1966), the function $d(z)$ is a measurable one.

2.4   The calculation of conditional expectations

To solve the equations (2.2.1) we must calculate the conditional expectation with regard to random vectors

$$z_n = (x_i, y_i, \ i = 1, \dots, n).$$

The methods for the calculation of such conditional expectation are well known if the points of the observations x_i are fixed, but not when the observation points x_{n+1} depend on z_n. However, it will be shown that under some conditions the conditional expectations in the case of dependent observations x_i can be calculated using the same formulae as in the case when the observations are fixed.

Suppose that there exists a continuous density function

$$p_{x_1 x_2}(y_1, y_2) = \lim_{\substack{\Delta_1 \to 0 \\ \Delta_2 \to 0}} (1/(\Delta_1 \Delta_2))$$

$$\times P\{\omega : y_1 \leq f(x_1, \omega) < y_1 + \Delta_1, y_2 \leq f(x_2, \omega) < y_2 + \Delta_2\}.$$

(2.4.1)

Assume that the decision function d_1' is continuous almost everywhere and that

$$x_2 = d_1'(x_1, f(x_1, \omega)), \ x_1 \in A.$$
(2.4.2)

Then

$$\lim_{\substack{\Delta_1 \to 0 \\ \Delta_2 \to 0}} (1/(\Delta_1 \Delta_2))$$

$$\times P\{\omega : y_1 \le f(x_1, \omega) < y_1 + \Delta_1, y_2 \le f(d'_1(x_1, f(x_1, \omega), \omega)) < y_2 + \Delta_2\}$$

$$= \lim_{\substack{\Delta_1 \to 0 \\ \Delta_2 \to 0}} (1/(\Delta_1 \Delta_2))$$

$$\times P\{\omega : y_1 \le f(x_1, \omega) < y_1 + \Delta_1, y_2 \le f(d'_1(x_1, y_1), \omega) < y_2 + \Delta_2\}$$

$$= \lim_{\substack{\Delta_1 \to 0 \\ \Delta_2 \to 0}} (1/(\Delta_1 \Delta_2))$$

$$\times P\{\omega : y_1 \le f(x_1, \omega) < y_1 + \Delta_1, y_2 \le f(x_2, \omega) < y_2 + \Delta_2\}$$

$$= p_{x_1 x_2}(y_1, y_2) \tag{2.4.3}$$

where (2.4.3) holds at the continuity points of the decision function d'_1. The set of such points has a probability 1 because the function d'_1 is continuous almost everywhere and there exists a density function which is continuous.

A generalization of relation (2.4.3) to the case of an n-dimensional density function is given by

THEOREM 2.4.1. *Suppose that there exists the continuous density function*

$$p_{x_1, \cdots, x_n}(y_1, \ldots, y_n)$$

$$= \lim_{\substack{\Delta_1 \to 0 \\ \cdots \\ \Delta_n \to 0}} (1/(\Delta_1, \ldots, \Delta_n))$$

$$\times P\{\omega : y_1 \le f(x_1, \omega) < y_1 + \Delta_1, \ldots, y_n \le f(x_n, \omega) < y_n + \Delta_n\}.$$

Assume that the decision functions d'_n, $n = 1, 2, \ldots$ are continuous almost everywhere and that

$$x_{n+1} = d'_n(x_1, f(x_1, \omega), d'_1(x_1, f(x_1, \omega)), f(d'_1(x_1, f(x_1, \omega)), \omega), \ldots) \tag{2.4.5}$$

Then

$$\lim_{\substack{\Delta_1 \to 0 \\ \cdots \cdots \cdots \\ \Delta_n \to 0}} (1/(\Delta_1, \dots, \Delta_n))$$

$$\times P\{\omega : y_1 \leq f(x_1, \omega) < y_1 + \Delta_1, \ y_2 \leq f(d_1{}'(x_1, f(x_1, \omega)), \omega)$$

$$< y_2 + \Delta_2, \dots, y_n \leq f(d'_{n-1}, \dots, \omega) < y_n + \Delta_n\}$$

$$= p_{x_1, \dots, x_n}(y_1, \dots, y_n) \tag{2.4.6}$$

with probability 1. The equality (2.4.6) for the n-dimensional case can be proved in exactly the same way as the equality (2.4.3) for the two-dimensional case.

2.5   The one-step approximation

The solution of recurrent equations (2.2.1) is difficult. So some approximation is indispensable. The one-step approximation, see Mockus (1972), when the next observation is considered as the last one, is simple and natural. In such a case

$$u'_N(z_N) = \inf_{x \in A} E\{f(x)/z_N\}$$

$$u'_{n-1}(z_{n-1}) = \inf_{x \in A} R_x^{n-1}, \ x_n \in \arg \inf_{x \in A} R_x^{n-1}, \ 1 \leq n \leq N \tag{2.5.1}$$

where

$$R_x^{n-1} = E\{\inf_{v \in A} E\{f(v)/z_{n-1}, x, f(x)\}|z_{n-1}\}. \tag{2.5.2}$$

Assume the existence of the conditional density of $f(v)$ with regard to z_{n-1} and denote it by

$$p_v(y/z_{n-1})$$

where y means a value of $f(v)$, $v \in A$.

Let

$$p_v^n(x, y') = p_v(y/z_{n-1}, x, y'),$$

$$c_{n-1} = \{\inf_{v \in A} \int_{-\infty}^{\infty} y\, p_v(y/z_{n-1})\, dy$$

and

$$c_n(x, y') = \inf_{v \in A} \int_{-\infty}^{\infty} y\, p_v^n(x, y')\, dy.$$

Suppose

$$c_n(x, y') = \min(c_{n-1}, y'). \tag{2.5.3}$$

Then from (2.5.2)

$$R_x^{n-1} = \int_{-\infty}^{\infty} c_n(x, y)\, p_x(y/z_{n-1})\, dy. \tag{2.5.4}$$

It follows from assumption (2.5.3) that

$$R_x^{n-1} = \int_{-\infty}^{\infty} \min(c_{n-1}, y)\, p_x(y/z_{n-1})\, dy$$

$$= c_{n-1} - \int_{-\infty}^{c_{n-1}} (c_{n-1} - y)\, p_x(y/z_{n-1})\, dy. \tag{2.5.5}$$

It was shown by Mockus (1978) that (2.5.1) is asymptotically optimal in the sense that it converges to a global minimum of any continuous function. The convergence conditions will be given later in 4.2. However, when n is not large enough, the algorithm (2.5.2) tends to seek the minimum in the neighbourhood of the best point, more or less neglecting the areas where the observed values where not so good. This occurs because the influence of the remaining observations $n + 2$, $n + 3, \ldots, N$ is neglected in the equations (2.5.1).

If we want to make the one-step Bayesian method (2.5.1) more 'global' and so more like (2.2.1) we should include the influence of the remaining observations $n + 2, n + 3, \ldots, N$ into the equations (2.5.1). This can be done as a first approximation by the substitution of C_N for c_n in (2.5.5).

$$C_N = c_n - \varepsilon_N. \tag{2.5.6}$$

If the results of the observations can be regarded as independent Gaussian with expectation μ and variance σ^2, it is reasonable to assume the following equality

$$\varepsilon_N = E\{c_n\} - E\{C_N\}. \tag{2.5.7}$$

Then from Kramer (1946)

$$E\{C_N\} = \mu - \sigma\sqrt{(2 \ln N)} + 0(\ln N/\ln \ln N) \tag{2.5.8}$$

and

$$E\{c_n\} = \mu - \sigma\sqrt{(2 \ln n)} + 0(\sqrt{(\ln n)} / \ln \ln n). \tag{2.5.8.1}$$

Disregarding the remainder terms in (2.5.8) and (2.5.8.1) ε_N can be expressed as

$$\varepsilon_N = \sigma\left(\sqrt{(2 \ln N)} - \sqrt{(2 \ln n)}\right). \tag{2.5.9}$$

Formula (2.5.9) was derived under the assumption of independent observations. They are actually dependent because they must satisfy (2.5.1). Hence (2.5.9) should be adjusted to the real data, for example, by multiplication of (2.5.9) by the factor α and the addition of the positive number ε, i.e.

$$\varepsilon_N' = \alpha\varepsilon_N + \varepsilon, \quad \alpha > 0, \quad \varepsilon > 0$$

where α, ε should be evaluated from the results of experimentation.

EXAMPLE 2.5.1. Suppose that it is possible to remember the results of only one observation and that after each observation we must decide which observation to remember: an old one or a new one. This is a special case of Bayesian methods with limited memory considered by Mockus (1972). Denote by x_{i_n} the point of the old observation (the one which remains after the n-th observation). Denote by x_{n+1} the point of the new observation. The results of the old and new observations will be denoted by y_{i_n} and y_{n+1}, respectively.

Assume that

$$\inf_x E\{f(v)/z_{i_n}\} = \min (y_{i_n}, \mu) \tag{2.5.10}$$

where μ is the *a priori* expectation and $z_{i_n} = (x_{i_n}, y_{i_n})$. This assumption is true, for example, for Gaussian functions $f(x)$ with non-negative correlation which approaches zero when the distance between the points is large and set A is unbounded.

Denote

$$u_n(z_{i_n}) = \min (y_{i_n}, \mu) \tag{2.5.11}$$

$$u'_{n-1}(z_{i_{n-1}}, z_n)$$

$$= \min_{a \in [0,1]} [aE\{u_n(z_{i_n})/(z_{i_{n-1}}, z_n)\} + (1-a) E\{u_n(z_n)/(z_{i_{n-1}}, z_n)\}]$$

$$\tag{2.5.12}$$

$$u_{n-1}(z_{i_{n-1}}) = \inf_x E\{u'_{n-1}(z_{i_{n-1}}, z_n)/z_{i_{n-1}}\}. \tag{2.5.13}$$

Value $a = 1$ means keep an old observation, value $a = 0$ means keep the new one.
From (2.5.11) and (2.5.12)

$$u'_{n-1}(z_{i_{n-1}}, z_n) = \min_a [a \min (y_{i_{n-1}}, \mu) + (1-a) \min (y_n, \mu)]$$

$$= \min (y_{i_{n-1}}, y_n, \mu). \tag{2.5.14}$$

This means that the best Bayesian decision is to keep the observation with the minimal observed values.
From (2.5.13) and (2.5.14)

$$u_{n-1}(z_{i_{n-1}}) = \inf_x E\{\min (y, y_{i_{n-1}}, \mu)/(x_{i_{n-1}}, y_{i_{n-1}})\}. \tag{2.5.15}$$

In the case of a homogeneous and isotropic Gaussian function where the expectation is μ, the standard variance is σ^2 and the correlation $\rho(x', x'')$ between the points x' and x'' depends only on the distance $\|x' - x''\|$

$$u_{n-1}(z_{i_{n-1}}) = \inf_\rho (\sqrt{(2\pi)}\sigma_n)^{-1} \int_{-\infty}^{\infty} \min (y, y_{i_{n-1}}, \mu) \exp (-1/2(\frac{y - \mu_n}{\sigma_n})^2)dy$$

$$\tag{2.5.16}$$

where the conditional expectation

$$\mu_n = \mu - \frac{\rho(x, x_{i_{n-1}})(y_{i_{n-1}} - \mu)\sigma}{\sigma_n} \tag{2.5.17}$$

and the conditional variance

$$\sigma_n^2 = \sigma^2(1 - \rho^2(x, x_{i_{n-1}})) \tag{2.5.18}$$

It follows from (2.5.16) that the optimal Bayesian decision is to keep the observations with minimal value and to make the next observation on the sphere of radius $r_n = \rho^{-1}(\rho_n) < \infty$ if $y_{i_{n-1}} < \mu$, where ρ_n minimizes (2.5.16).

So example 2.5.1 explains, in Bayesian terms, the meaning of the well known method of random search by Rastrigin (1968).

2.6. The adaptive Bayesian approach

In the definition of Bayesian optimality (2.1.11) it was assumed that the *a priori* probability P is completely known before observations are started. However, it would be more realistic to assume that at least some parameters θ of P are unknown and can be estimated only on the base of observed values. The distributions G of θ can also be unclear. This means that the *a priori* probability P is not fixed and can change under the influence of the observed results and so

$$P = P_n, \ n = 0, 1, \dots N. \tag{2.6.1}$$

In such a case the condition of the Bayesian optimality

$$\inf_{d \in D} \int_\Omega \delta(d, \omega) \, P_n(d\omega), \ n = 0, 1, \dots, N \tag{2.6.2}$$

is not well-defined since P_n depends on n.

A way to give the condition (2.6.2) meaning is the one-step approximation

$$\inf_{d_n \in D_n} \int_\Omega \delta(d, \omega) \, P_n(d\omega) \tag{2.6.3}$$

where each component d_n of the decision function d is chosen under the assumption that the distribution P_n will not change later.

The consistency conditions of Kolmogorov for the distributions P_n in a case (2.6.5), when P_n depends on n, no longer seem necessary since after each n we are, strictly speaking, considering different stochastic functions which represent different stochastic models of the functions to be minimized. The change in P_n means that we are updating the corresponding statistical models on the basis of the results of the observations.

Some other conditions must be defined to provide consistency of different statistical models if the conditions of Kolmogorov are omitted. Most natural (in the

Bayesian case) is the condition of continuity of the risk function. Under this assumption the much simpler version of the Bayesian method will be developed in 5.2. It will be called the adaptive Bayesian method because it 'adapts' more directly to the new observed data by changing the statistical models P_n. The classical Bayesian approach adapts indirectly to the observed results by means of updating the *a posteriori* distributions in the framework of a fixed stochastic model, the so-called *a priori* distribution.

THE AXIOMATIC NON-PROBABILISTIC JUSTIFICATION OF BAYESIAN OPTIMALITY CONDITIONS

3.1 Introduction

A definition of Bayesian optimality (2.1.11) was derived assuming that the following conditions were satisfied:

3.1.1. An optimal method should minimize the average losses of deviation.

3.1.2. The losses connected with the deviation from the global minimum f_0 are a linear function of the difference $f(x_{N+1}) - f_0$, where x_{N+1} is the point of final decision.

3.1.3. The function to be minimized is a sample of some stochastic function defined by the *a priori* probability measure P.

Those conditions will be derived from some simple and clear assumptions later in this chapter.

3.2 The linearity of the loss function

The losses which occur when the method of optimization d is applied to the sample of $f(x)$ defined by the index ω, will be denoted by $l(d, \omega)$. Then the average losses can be expressed, see De Groot (1970), by an integral (if it exists) of the loss function $l(d, \omega)$

$$L(d) = \int_{\Omega} l(d, \omega) P(d\omega). \tag{3.2.1}$$

The definition of the best method in the sense of the average deviation (3.2.1) is correct if the loss function $l(d, \omega)$ defines such ordering of methods that

$$d' \geqslant d'' \tag{3.2.2}$$

if and only if

$$L(d') \leq L(d'').$$ (3.2.2)

The symbol $\gtrsim$ in the expression (3.2.2) means that the method d' is at least as good as the method d''.

The deviation from the global minimum can be defined either in the space of variables $x \in R^m$ or in the space of values of the function $f(x) \in R$. In the case of the first definition the loss function $l(d, \omega)$ would be of m-dimensional argument. The first definition is also inconvenient when the global minimum is not unique (there exist several points $x \in A$ where the minimal value of $f(x)$ is attained) because the distance $\|x - x_0\|$ from the point of global minimum x_0 is not uniquely defined. Therefore, in this book, the deviation from the global minimum will be defined on the space of the values of the function $f(x)$. This has better correspondence to the meaning of $f(x)$ since we should minimize the objective function $f(x)$ but not the distance $\|x - x_0\|$ of x from the point of the global minimum x_0. In this case the loss function can be expressed as the function of δ

$$l(d, \omega) = l'(\delta)$$ (3.2.4)

where

$$\delta = \delta(d, \omega) = f(x_{N+1}, \omega) - f(x_0, \omega) \in [0, \infty).$$ (3.2.5)

Here l' does not depend on d and ω when δ is fixed.

From (3.2.1) and (3.2.4)

$$L(d) = \int_\Omega l(d, \omega) P(d\omega) = \int_0^\infty l'(\delta) P'(d\delta),$$ (3.2.6)

where the probaility measure P' is defined by the equality

$$P'\{\delta \in B\} = P\{\omega: \delta(d, \omega) \in B\}$$ (3.2.7)

for all Borel sets $B \in R$.

Let us consider the method $d \in D$ as the point of the linear space and suppose that the following assumptions hold:

ASSUMPTION 3.2.8. The set of methods D is convex. This means that if $d' \in D$ and $d'' \in D$ then $\alpha d' + (1 - \alpha) d'' \in D$, $\alpha \in [0, 1]$.

The expression

$$\alpha d' + (1 - \alpha)d''$$

can be regarded as a lottery when method d' is used with probability α and method d'' is used with probability $(1 - \alpha)$.

ASSUMPTION 3.2.9. The set D is completely ordered by the relation $\gtrsim$. This means that the relation $d' \gtrsim d''$ or the relation $d'' \gtrsim d'$, or both, namely, the relation $\sim$, holds for all d', $d'' \in D$.

ASSUMPTION 3.2.10. The ordering $\gtrsim$ is continuous. This means that the sets $\{\alpha : \alpha d' + (1 - \alpha)d'' \geq d'''\}$ and $\{\alpha : d''' \geq \alpha d' + (1 - \alpha)d''\}$ are closed for all d', d'', $d''' \in D$.

ASSUMPTION 3.2.11. The lotteries composed of indifferent methods are indifferent. This means that if d', $d'' \in D$ and $d' \sim d''$ then for all $d''' \in D$ and all $\alpha \in [0, 1]$

$$\{\alpha d' + (1 - \alpha)d'''\} \sim \{\alpha d'' + (1 - \alpha)d'''\}.$$

THEOREM 3.2.1. *Suppose that the assumptions 3.2.8 to 3.2.11 hold. Then there exists a linear functional $L(d)$ such that for all d', $d'' \in D$ the relation*

$$d' \gtrsim d'' \tag{3.2.12}$$

holds if and only if

$$L(d') \leq L(d''). \tag{3.2.13}$$

   If there exists another linear functional $L'(d)$ satisfying conditions (3.2.12) and (3.2.13) for all d', $d'' \in D$, then

$$L'(d) = aL(d) + b \tag{3.2.14}$$

where $a > 0$.

The proof is given by Herstein and Milnor (1953).

The linear functional $L(d)$ in case (3.2.4) and (3.2.5) is defined as an integral

$$L(d) = \int_0^\infty l'(\delta)\, P'(d\delta) \tag{3.2.15}$$

where l' is a linear function of δ. This means that the linearity of the loss function (see condition 3.1.2) follows from assumptions 3.2.8 to 3.2.11 of Theorem 3.2.1 and conditions (3.2.4) and (3.2.5).

3.3 The existence of the unique *a priori* probability corresponding to subjective preferences

The assumption 3.1.3 that the function $f(x)$ to be minimized is a sample of some stochastic function is true in some cases of engineering design and planning.

However, the function $f(x)$ is more often defined by unique conditions which cannot be repeated in the future. One of the reasons is that the need for the designing of a new system usually arises only when technological, economic, social and environmental conditions change.

The apparent impossibility of defining the *a priori* probability P in the case of unique, unrepeatable conditions is generally supposed to be the main disadvantage of the Bayesian approach. However, it was shown by Katkauskaite and Zilinskas (1977) that under some assumptions there exists the *a priori* probability P which defines average losses $L(d)$ and corresponds to the relation of subjective preferences $\succeq$ in the sense that

$$B_i \succeq B_j \tag{3.3.1}$$

if and only if

$$P(B_i) \geq P(B_j). \tag{3.3.2}$$

Let us consider these assumptions.

It is assumed that for a fixed $x_j \in A, j = 1, \ldots, l,\ l = 1, 2, \ldots$ a subjective likelihood relation is defined on the set of events B_i where the event B_i means that the vector $(f(x_1), \ldots f(x_l))$ belongs to the l-dimensional interval

$$B_i = \mathop{\times}_{k=1}^{l} B_i^k \tag{3.3.3}$$

where $\times$ denotes the Cartesian product and event B_i^k means that $f(x_k) \in B_i^k$. Here

$$B_i^k = (a_k^i, b_k^i]\ \text{if}\ b_k^i\ \text{is finite and}\ B_i^k = (a_k^i, b_k^i)\ \text{if it is not.} \tag{3.3.4}$$

B_i can also be an empty set $a_k^i \le b_k^i$, $i = 1, \dots , m, k = 1, \dots , l$.

In the well known papers (see De Groot (1970), Fine (1973), Savage (1954)) the likelihood relation is usually assumed to be defined not only on intervals, but also on an algebra of intervals. This makes the testing of preferences by psychological experimentation more complicated.

A pair of intervals B_i, B_j will be called an s-pair if

$$a_k^i = a_k^j \text{ and } b_k^i = b_k^j, \text{ when } k \ne s$$

$$a_k^i = a_k^j \text{ or } b_k^i = b_k^j, \text{ or } a_k^i = b_k^j \text{ or } a_k^j = b_k^i, \text{ when } k = s$$

An s-pair will be called a lower (upper) s-pair if

$$a_s^i = a_s^j \; (b_s^{\ i} = b_s^j).$$

A set of intervals B_i, $i = 1, \dots , l$ will be called an s-system if

$$a_k^i = a_k^j \text{ and } b_k^i = b_k^j \text{ if } k \ne s$$

It is obvious that the union if s-pairs is an interval.

It is assumed that the relation $\succeq$ satisfies five conditions:

ASSUMPTION 3.3.5. It is a complete ordering on a set of all intervals B_i and an empty set $\varnothing$.

ASSUMPTION 3.3.6. If $B_1 \cap B_2 = B_3 \cap B_4 = \varnothing$, where B_1, B_2 and B_3, B_4 are s-pairs for some $s = 1, \dots , l$, then from $B_1 \succeq B_3, B_2 \succeq B_4$ it follows that $B_1 \cup B_2 \succeq B_3 \cup B_4$. If one of the first two relations is strict $\succ$, then the last relation should be strict $\succ$.

ASSUMPTION 3.3.7. $B \succ \varnothing$ iff $\mu(B) > 0$, $B \cup A \sim B$ if $\mu(A) = 0$, $\mu(A)$ is a Lebesgue measure of A.

ASSUMPTION 3.3.8. If $B_1 \succ B_2 \succ \varnothing$, then for any $s = 1, \dots , l$ there exists B_3, B_4 such that $B_3 \subset B_1, B_4 \subset B_1, B_2 \sim B_3 \sim B_4$, where B_3, B_1 is a lower s-pair, and B_4, B_1 is an upper s-pair. Here $\sim$ denotes the equivalence relation.

ASSUMPTION 3.3.9. If $B_i \sim C_i$ and $B_i \cap B_j = \varnothing$, $C_j \cap C_j = \varnothing$, $i = j$, then

$$\bigcup_{i=1}^{n} B_i \sim \bigcup_{i=1}^{n} C_i$$

Condition 3.3.5 appears rather clear and natural. Condition 3.3.6 can be regarded as an independence from irrelevant events; conditions 3.3.7 and 3.3.8 mean a sort of continuity of relation $\succcurlyeq$; condition 3.3.9 defines the extension of the relation $\succcurlyeq$ from intervals to an algebra of intervals.

Suppose $B_j, j = 1, \dots, k$ is an s-system such that

$$B_j \cup B_i = \varnothing, \; j \neq i$$

for any pair i, $j = 1, \dots, k$.

Denote event $f(x_n) \in (-\infty, \infty)$ as R or B_∞^n.
Let

$$B = \bigcup_{j=1}^{k} B_j \; \text{ and } \; B_j \subset \underset{i=1}{\overset{l}{\times}} B_\infty^i = R^l.$$

Condition 3.3.8 shows that there exist real numbers $c_l^1, \dots, c_l^k$ such that

$$B_1 \sim R^{l-1} \times (-\infty, c_{l}^1],$$

and

$$B_2 \sim R^{l-1} \times (c_{l}^1, c_l^2],$$

and so on until

$$B_k \sim R^{l-1} \times (c_l^{k-1}, c_l^k].$$

It follows from condition 3.3.9 that there exists such c_l^k that

$$B \sim R^{l-1} \times (-\infty, c_l^k].$$

Then from 3.3.8 it follows that for any s, $1 \leq s \leq l$ there exists c_s^k such that

$$B \sim R^{l-1} \times (-\infty, c_s^k], \; \text{ if } \mu(R^\backslash B) > 0$$

and

$$B \sim R^{l-1} \times (-\infty, \infty) = R^l, \; \text{ if } \mu(R^\backslash B) = 0. \tag{3.3.9.1}$$

Condition (3.3.9.1) shows how the relation $\succcurlyeq$ can be extended from intervals to finite unions of intervals.

THEOREM 3.3.1. *If assumptions 3.3.5 to 3.3.9 hold, then there exists a unique l-dimensional probability density function*

$$p_{x_1,\dots,x_l}(y_1, \dots, y_l), \quad l = 1, 2, \dots \tag{3.3.10}$$

which corresponds to the subjective likelihood relation in the sense that

$$B_i \succcurlyeq B_j \tag{3.3.11}$$

if and only if

$$P(B_i) \geqslant P(B_j) \tag{3.3.12}$$

where

$$P(B) = \int_B p_{x_1,\dots,x_l}(y_1, \dots, y_l)\, dy_1, \dots, dy_l, \tag{3.3.13}$$

Proof. The following seven conditions hold on an algebra of finite unions of intervals when the relation $\succcurlyeq$ is extended by equality (3.3.9.1)

3.3.15. $R^l \succ \varnothing$

3.3.16. $B_1 \succcurlyeq B_2$ or $B_2 \succcurlyeq B_1$ for any B_1, B_2

3.3.17. If $B_1 \succcurlyeq B_2$ and $B_2 \succcurlyeq B_3$ then $B_1 \succcurlyeq B_3$

3.3.18. $B \succcurlyeq \varnothing$

3.3.19. When $B_1 \cap (B_2 \cup B_3) = \varnothing$, then $B_2 \succcurlyeq B_3$ iff $B_1 \cup B_2 \succcurlyeq B_1 \cup B_3$

3.3.20. Topological space generated by an algebra of intervals has a countable base.

3.3.21. For any finite k and $s = 1, \dots, l$ there exists an s-system such that $i \leq k < \infty$ and

$$R^l = \bigcup_{j=1}^{k} B_j, \quad a_s^l = -\infty, \quad b_s^k = \infty$$

$$B_j \cap B_{j+1} = \varnothing, \; B_j \sim B_{j+1}, \; j = 1, \dots, k-2, \; B_k \lesseqgtr B_1,$$

B_i and B_j is an s-pair for any $i, j = 1, \dots, k$.

Conditions 3.3.15, 3.3.18 and 3.3.19 are the consequences of 3.3.7 and 3.3.9; conditions 3.3.16 and 3.3.17 follow from 3.3.5 and 3.3.9. Condition 3.3.20 is well known (see Kolmogorov and Fomin (1968)). Now we shall prove condition 3.3.21. Let us partition R^l

$$R^l = B_1 \cup B_2$$

where both B_1 and B_2 are finite unions of the intervals. Then from (3.3.9.1) it follows that for any $s = 1, \dots, l$

$$B_1 \sim R^{l-1} \times \bigcup_{j=1}^{k-1} (c_s^{j-1}, c_s^j] \text{ and } B_2 \sim R^{l-1} \times (c_s^{k-1}, c_s^k)$$

where $c_s^0 = -\infty, c_s^k = \infty$.

From here

$$R^l = R^{l-1} \times \bigcup_{j=1}^{k-1} (c_s^{j-1}, c_s^j] \cup (c_s^{k-1}, c_s^k) \qquad (3.3.22)$$

This means that condition (3.3.21) is true for any i if there exists a partition of an interval $(-\infty, \infty)$ into $k, i \le k < \infty$ subintervals such that

$$(-\infty, \infty) = \bigcup_{j=1}^{k-1} (c^{j-1}, c^j] \cup (c^{k-1}, c^k) \qquad (3.3.23)$$

where $c^0 = -\infty, c^k = \infty$

$$(c^{j-1}, c^j] \sim (c^j, c^{j+1}], \; j = 1, \dots, k-2$$

$$(c^{k-1}, c^k) \lesseqgtr (c^0, c^1].$$

We begin the proof of (3.3.23) by showing that for each interval $(a, b]$ there exists a partition of $(-\infty, \infty)$ such that

$$(-\infty, \infty) = \bigcup_{j=1}^{k-1} (c^{j-1}, c^j] \cup (c^{k-1}, c^k), \; k < \infty, c^0 = -\infty, c^k = \infty \qquad (3.3.23.1)$$

$$(c^{j-1}, c^j] \sim (a, b], \ j = 1, \dots, k-1 \tag{3.3.23.2}$$

$$(c^{k-1}, c^k] \lessgtr (a, b] \tag{3.3.23.3}$$

Since $(a, b] \prec (-\infty, \infty)$, then from 3.3.8 it follows that there exists c^1 such that $(-\infty, c^1) \sim (a, b]$. If $(c^1, \infty) \lessgtr (a, b]$, then $k = 2$ and the proof of (3.3.23) is completed. Otherwise, there exists c^2 such that $(c^1, c^2] \sim (a, b]$, etc.

Suppose that c^j is an infinite sequence. Let $c^j \to c_0 < \infty$. (The case $c^j \to \infty$ can be considered in a similar way). From 3.3.8 it follows that there exists $c, -\infty < c < c_0$, such that $(c, c_0] \sim (a, b]$.

Since $(c^{j-1}, c^j] \sim (a, b]$ for each j, then $(c^{j-1}, c^{j+1}] \succ (a, b]$.

Nevertheless the convergence of c^j to c_0 gives the inequality $c < c^j < c_0$ for $j \geq L$ and some L. Consequently $(c^L, c^{L-3}] \subset (c, c_0]$ and $(c^L, c^{L+2}] \succ (c, c_0]$. The contradiction proves (3.3.23).

Let i be given, and $(a^1, b^1]$ be an arbitrary finite interval. Let $(-\infty, \infty)$ be partitioned into subintervals in accordance with (3.3.23), (3.3.23.1) and (3.3.23.2) with $(a^1, b^1]$. If the number of subintervals k_1 is greater than i, it is the desired result. Otherwise consider the intervals $(a^1, d^1], (d^1, b^1]$ where $d^1 = (a^1 + b^1)/2$. We denote the interval which is minimal in the sense of the likelihood relation by $(a^2, b^2]$ and let $(-\infty, \infty)$ be partitioned once more by (3.3.23) for $(a^2, b^2]$. Now the number of subintervals will be $k_2 > 2k_1$, if $k_2 \geq i$, then the proof is completed. If it is not then consider the intervals $(a^2, d^2], (d^2, b^2]$ where $d^2 = (a^2 + b^2)/2$ etc. After a finite number of such steps we shall achieve the appropriate partition.

Since the relation $\succcurlyeq$ corresponds to the conditions (3.3.15) to (3.3.21) then it follows, see Fine (1973), that there exists the unique function P, defined on finite unions of the intervals $B_i \subset R^m$ such that

1) $P(b_j) \geq P(B_i)$ iff $B_j \succcurlyeq B_i$, \hfill (3.3.24)

2) $P(R^l) = 1, \ P(\emptyset) = 0$, \hfill (3.3.25)

3) If $B_j \cap B_i = \emptyset, \ i \neq j$, then

$$P(\bigcup_{i=1}^{k} B_i) = \sum_{i=1}^{k} P(B_i) \tag{3.3.26}$$

It follows from 3.3.21 that $P(B)$ is continuous in the sense that if $B_j \supset B_{j-1}$ and $\bigcap\limits_{j=1}^{\infty} B_j = \varnothing$ then

$$\lim_{j \to \infty} P(B_j) = 0. \tag{3.3.27}$$

Suppose on the contrary

$$\lim_{j \to \infty} P(B_j) = \varepsilon > 0.$$

Then from 3.3.21 it follows that there exists an m-system $A_j, j = 1, \ldots, k$ such that

$$R^m = \bigcup\limits_{i=1}^{k} A_i \quad \text{and} \quad P(A_i) < \varepsilon/2.$$

We can have an equivalent representation of A_i such that

$$A_i \sim R^{l-1} \times (a_i^{i-1}, a_i^i]$$

since

$$P(R^l \times (a_i^{i-1}, a_i^i)) < \varepsilon/2.$$

We can also have an equivalent representation of B_j

$$B_j \sim R^{l-1} \times (b_l^{j-1}, b_l^j].$$

From here and from the assumption that $B_j \supset B_{j-1}$ and $\bigcap\limits_{j=1}^{\infty} B_j = \varnothing$ it follows that

$$(b_l^{j-1}, b_l^j] \supset (b_l^j, b_l^{j+1}] \quad \text{and} \quad \bigcap\limits_{j=1}^{\infty} (b_l^j, b_l^{j+1}] = \varnothing.$$

Then

$$\lim_{j \to \infty} b_l^{j-1} = \lim_{j \to \infty} b_l^j$$

and there exist M and j such that

$$(b_l^{j-1}, b_l^j] \subset (a_i^{i-1}, a_i^i], \ i \geq M. \tag{3.3.28}$$

However

$$P(R^{l-1} \times (b_l^{j-1}, b_l^j]) \geq \varepsilon \text{ and } P(R^{l-1} \times (a_l^{i-1}, a_l^i]) < \varepsilon \qquad (3.3.29)$$

which contradicts (3.3.28).

It is well known that from the continuity of P, condition (3.3.27), and conditions (3.3.24) to (3.3.26) it follows that there exists a unique expansion of P to a Borel field, see Neveu (1969). Condition 3.3.7 means that P is absolutely continuous with respect to the Lebesgue measure, and so there exists the unique l-dimensional probability density function corresponding to the relation $\succeq$.

The theorem similar to Theorem 3.3.1 was proved by Katkauskaite and Zilinskas (1977). The difference lies in the formulation of the extension assumption. Katkauskaite and Zilinskas (1977) assumed the condition (3.3.9.1) instead of (3.3.9).

Theorem 3.3.1 does not say anything about the consistency of l-dimensional density functions in Kolmogorov's sense. Theorem 3.3.1 was formulated in such a way because it was shown in section 2.6 that the consistency is not necessary when *a priori* distributions depend on n, because then we have not one stochastic function but many.

If we wish to regard the function $f(x)$ as a sample path of some stochastic function, then it is necessary to set some additional consistency conditions.

3.3.30. $B_1 \times \ldots \times B_n \times B_\infty^{n+1} \sim B_1 \times \ldots \times B_n$.

Here : $B_\infty^{n+1} = (-\infty, \infty)$.

3.3.31. $B_1 \times \ldots \times B_n \sim B_{i_1} \times \ldots \times B_{i_n}$.

Here : $\{i_k\} = \{1, \ldots, n\}$.

THEOREM 3.3.2. *If the relation $\succeq$ satisfies conditions 3.3.30 and 3.3.31 then the functions* $p_{x_1, \ldots, x_l}(y_1, \ldots, y_l)$, $l = 1, 2, \ldots$ *defined in Theorem 3.3.1 are a consistent family of l-dimensional probability density functions.*

COROLLARY 3.3.3. *If assumptions 3.3.5 to 3.3.9 and 3.3.30 and 3.3.31 hold, then there exists a stochastic function, finite dimensional distributions of which correspond to the relation $\succeq$.*

Corollary 3.3.3 means that the function $f(x)$ can be regarded as a sample path of some stochastic function defined by a consistent family of l-dimensional distribution functions (3.3.10) and (3.3.13).

The proof of Theorem 3.3.2 was given by Zilinskas (1978) for the one-dimensional case $A \in R$ and with a slightly different formulation of condition (3.3.9).

The extension to the multi-dimensional case $A \in R^m$ was done by Katkauskaite and Zilinskas (1977) using the same framework of proof as in the one-dimensional case.

EXAMPLE 3.3.1. Complete indifference.

Suppose that we have no preference for any B_i. This means that

$$B_i \sim B_j \text{ for all non-empty } B_i, B_j. \tag{3.3.34}$$

This assumption violates the last part of condition 3.3.6.

Suppose that

$$B_1 \cap B_2 = B_3 \cap B_4 = \emptyset, B_1, B_2, B_4 \neq \emptyset, B_3 = \emptyset, \text{ and } B_1 \cup B_2 \text{ is an}$$
interval.

Then from (3.3.14)

$$B_1 \sim B_2 \sim B_4,$$

and from (3.3.7)

$$B_1 \succcurlyeq B_3.$$

Then from the last part of condition 3.3.6 it follows that

$$B_1 \cup B_2 \succ B_3 \cup B_4$$

Since B_3 is an empty set, $B_3 \cup B_4 = B_4$ and $B_1 \cup B_2 \succ B_4$, which contradicts the assumption (3.3.34).

EXAMPLE 3.3.2. Complete uniformity.

Suppose that

$B_i \geqslant B_j$, if $v(B_i) \geqslant v(B_j)$ (3.3.35)

$B_i \sim B_j$, if neither $v(B_i) < v(B_j)$ nor $v(B_j) < v(B_i)$,

where

$v(B_i) = b_i - a_i$,

$B_i = (a_i \, b_i]$,

$v(\varnothing) = 0$.

Denote

$B_1 = (-\infty, a]$

$B_2 = (a, \infty)$

$B_3 = (-\infty, \infty)$

$B_4 = \varnothing$.

Here

$B_1 \cap B_2 = B_3 \cap B_4 = \varnothing$,

$B_1 \sim B_3$

because $v(B_1) = \infty$, $v(B_3) = \infty$ and $B_2 > B_4$.

It follows that

$B_1 \cup B_2 \sim B_3 \cup B_4$ (3.3.36)

because $B_1 \cup B_2 = B_3$ and $B_3 \cup B_4 = B_3$.

Relation (3.3.6) violates the last part of the condition (3.3.6).

EXAMPLE 3.3.3. Regional Uniformity.

Suppose that condition (3.3.15) holds for $B_i \subset B$, where $B \subset R^l$, for example $B = (0, 1]^l$. Here all the conditions hold and there exists the unique *a priori* probability density

$$p_{x_1, \ldots, x_l} (y_1, \ldots, y_l) = \prod_{k=1}^{l} p_{x_k}(y_k) \tag{3.3.37}$$

where

$$p_{x_k}(y_k) = b_k - a_k, \quad k = 1, \ldots, l. \tag{3.3.38}$$

The sample paths of the stochastic function $f(x)$ corresponding to (3.3.37) and (3.3.38) are very irregular and not necessarily measurable as functions of x.

3.4 Optimal method under uncertainty

Sometimes it is convenient to regard the problem of optimization as a problem of decision making under uncertainty, see De Groot (1962).

ASSUMPTION 3.4.1. For any fixed $\omega \in \Omega$ the relation $d' \geqslant d''$ holds if and only if

$$f(x_{N+1}(d'), \omega) \leq f(x_{N+1}(d''), \omega) \tag{3.4.1}$$

where $x_{N+1}(d)$ is the final decision achieved by the application of method d.

THEOREM 3.4.1. *If the assumptions 3.2.8 to 3.2.11, 3.3.5 to 3.3.9 and 3.4.1 hold, then the problem of optimization under uncertainty can be reduced to the problem of Bayesian optimization, in a sense that*

$$d' \geqslant d'' \tag{3.4.2}$$

if and only if

$$L(d') \leq L(d'') \tag{3.4.3}$$

where

$$L(d) = \int_{\Omega} f(x_{N+1}(d), \omega) \, P(d\omega),$$

and $x_{N+1}(d)$ is the final decision achieved by the application of method d and P is a probability measure.

Proof. It follows from Theorem 3.2.1 and assumptions 3.4.1 and 3.2.8 to 3.2.11 that there exists a linear functional $L(d)$ such that

$$d' \geqslant d'', \text{ iff } L(d') \leq L(d'') \tag{3.4.4}$$

The existence of the corresponding P follows from Theorem 3.3.1 and assumptions 3.3.5 to 3.3.9.

Theorem 3.4.1 means that in case of uncertainty the optimal method should minimize the average deviation (condition 3.1.1).

3.5   Non-linear loss functions

In some methods of global optimization based on stochastic models losses are represented as a step function

$$l(y) = I(y - y_0), \tag{3.5.1}$$

where

$$I(y) = \begin{cases} 0, & y < 0, \\ 1, & y \geq 0, \end{cases}$$

$$y = f(x_{N+1}(d), \omega),$$

and

y_0 is some fixed critical level.

Such loss functions are used by Kushner (1964), Strongin (1978) and Zilinskas (1982).The following conditions were assumed by Zilinskas (1980) to justify the use of a step function (3.5.1).

ASSUMPTION 3.5.1. The Gaussian conditional density functions with conditional mean μ and conditional variance σ^2 are completely ordered by the preference relation defined on a set of pairs (μ, σ).

Let y_0 be a desirable level of the objective function $f(x)$.

ASSUMPTION 3.5.2. When $\mu' < \mu''$, $\mu'' > y_0$, $\sigma' > 0$ then there exists σ such that the relation

$$(\mu', \sigma') > (\mu'', \sigma'')$$

is true, if $\sigma'' \leq \sigma$.

ASSUMPTION 3.5.3. When

$$\mu'' > y_0$$

then the relation

$$(\mu', \sigma') \succ (\mu'', 0)$$

is true for any μ' and $\sigma' > 0$.

ASSUMPTION 3.5.4. There exists μ', $(y_0 < \mu' < \mu'')$ such that

$$(\mu', \sigma') \succ (\mu'', \sigma'')$$

for any $\sigma' > 0$, $\mu'' > y_0$, $\sigma'' \geq 0$.

ASSUMPTION 3.5.5. The loss function $l(y, y_0)$ is continuous from the left.

Assumption 3.5.2 means that the larger mean will not be preferred if the variance is not large enough. Assumption 3.5.3 means that if the mean with zero variance exceeds the desirable level y_0 it will not be preferred.

THEOREM 3.5.1. *The unique (to the scale factor) loss function satisfying assumptions 3.5.1 to 3.5.5 is a step function*

$$l(y, y_0) = I(y - y_0) \tag{3.5.6}$$

where

$$I(y - y_0) = \begin{cases} 0, & \text{if } y \leq y_0, \\ 1, & \text{if } y > y_0 \end{cases}$$

and

$$(\mu', \sigma') \succ (\mu'', \sigma'')$$

if and only if

$$\int_{-\infty}^{\infty} l(y, y_0) \, n(y \,|\, \mu', \sigma') \, dy \; < \; \int_{-\infty}^{\infty} l(y, y_0) \, n(y \,|\, \mu'', \sigma'') \; dy.$$

Here $n(y \,|\, \mu, \sigma)$ means Gaussian density function with mean μ and variance σ^2.
The proof of Theorem 3.5.1 is given by Zilinskas (1985).

Let us test the compatibility of assumption 3.5.4 with the condition of continuity of relation $\succ$ in the sense that for any $\beta > 0$ there exists $\alpha(\beta) > 0$ such that

$$(\mu'' - \alpha(\beta), \sigma') \; \sim \; (\mu'', \sigma' + \beta), \tag{3.5.7}$$

$$(\mu'' - \alpha', \sigma') \; \succ \; (\mu'', \sigma' + \beta), \; \text{if } \alpha' > \alpha(\beta), \tag{3.5.8}$$

and

$$(\mu'' - \alpha'', \sigma') \; \prec \; (\mu'', \sigma' + \beta), \; \text{if } \alpha'' < \alpha(\beta). \tag{3.5.9}$$

The continuity conditions (3.5.7) to (3.5.9) mean that if $\alpha(\beta) < \mu'' - y_0$ then assumption 3.5.4 is compatible with the continuity of relation $\succ$ but if $\alpha(\beta) \geq \mu'' - y_0$ then it is not compatible.

Supposing that $\alpha(\beta)$ is an unbounded increasing function of β it follows that assumption 3.5.4 contradicts the continuity of relation $\succ$ if β is sufficiently large and consequently $\alpha(\beta) \geq \mu'' - y_0$.

The author prefers the continuity of relation $\succ$, so in this book, the linear loss function is used to define the Bayesian methods.

CHAPTER 4

STOCHASTIC MODELS

4.1 Introduction

The convenient way to define the method of optimization which minimizes the average deviation is to design the stochastic model of the function $f(x)$. In many cases the function $f(x)$ is determined by unique conditions which will not be repeated and so cannot be regarded as a stochastic function in the classical sense. Nevertheless such functions can be considered in the framework of some stochastic model if the conditions are satisfied concerning the relations of subjective likelihood. In the previous chapter the conditions were given for the case when there exists a family of finite-dimensional distribution functions which correspond to the relations of subjective likelihood.

If in addition the consistency conditions of Kolmogorov also hold, then the objective function can be regarded as a sample of some stochastic function. If not, then it can still be considered as a sample which is common for a sequence of different stochastic functions which can depend on the observed data and are defined by the corresponding sequence of finite-dimensional distribution functions, not necessarily consistent in the Kolmogorov sense.

The classical consistent stochastic model is more convenient for theoretical consideration because such important properties as continuity of sample functions can be defined by simple and clear conditions, see for example section 4.4. Unfortunately, the calculation of conditional probabilities in non-independent cases requires some inversion and multiplication of matrices of order N which is possible only if the number of observations N is not too large. Otherwise the Kolmogorov consistency should be changed to some weaker conditions, which also provide a sort of consistency of the statistical model but are simpler for computation, see section 5.2.

In this chapter mainly classical models will be considered.

An important problem is that in some cases it can be impossible to define the values $f(x)$ directly because only the sum

$$h(x) = f(x) + g(x)$$

can be observed. This will be called the case with noise

$$g(x) = g(x, \omega).$$

Here the noise $g(x)$ is supposed to be a stochastic function independent of $f(x)$ with zero expectation and bounded variance. If the expectation of the noise $g(x)$ is known, but not zero, then the problem can be easily reduced to the zero noise one. However if the expectation of the noise $g(x)$ is unknown, its influence cannot be eliminated and the minimization of $f(x)$ becomes meaningless in that sense.

4.2 Sufficient convergence conditions

In most practical applications the *a priori* distribution P cannot be precisely defined. Thus it would be very desirable to define a family of *a priori* distributions such that the Bayesian methods would converge to a global minimum of any continuous function. In this book we shall restrict ourselves to one-step Bayesian methods. We shall also consider the case when the value of $f(x)$ cannot be directly defined, i.e. when only the sum $h(x) = f(x) + g(x)$ can be observed, where $g(x)$ is noise.

Let us define the sets

$$\Omega_s^l = \{\omega : f(x_i, \omega) < s_i, i = 1, \dots, l\}$$

$$\Omega_v^l = \{\omega : g(x_i, \omega) < v_i, i = 1, \dots, l\}$$

$$\Omega_y^l = \{\omega : h(x_i, \omega) < y_i, i = 1, \dots, l\}$$

$$l = 1, 2, \dots \qquad (4.2.1)$$

Here x_i are fixed points from A, not necessarily the observation points. It is assumed that probabilities P of the sets Ω_s^l, Ω_v^l, Ω_y^l are defined by the l-dimensional distribution function F

$$F_{x_1, \dots, x_l}^s (s_1, \dots, s_l) = P\{\Omega_s^l\}$$

$$F_{x_1, \dots, x_l}^v (v_1, \dots, v_l) = P\{\Omega_v^l\}$$

$$F_{x_1, \dots, x_l}^y (y_1, \dots, y_l) = P\{\Omega_y^l\} \qquad (4.2.2)$$

Assume that there exist the corresponding probability density functions

$$p_{x_1, \dots, x_l}^s (s_1, \dots, s_l)$$

$$p^v_{x_1, \ldots, x_l} (v_1, \ldots, v_l)$$

$$p^y_{x_1, \ldots, x_l} (y_1, \ldots, y_l).$$ (4.2.3)

The conditional probability density with regard to the vector

$$z_n = (x_i, y_i, i = 1, \ldots, n)$$ (4.2.4)

will be denoted

$$p_x = p^n_x = p_x(s/z_n)$$ (4.2.5)

and defined uniquely as the usual relation of corresponding multi-dimensional probability density functions.

The conditional expectation can be expressed as

$$\mu^n_x = \mu_x(z_n) = \int_{-\infty}^{\infty} s\, p^n_x\, ds.$$ (4.2.6)

and the conditional variance as

$$\sigma^n_{xx} = {\sigma^n_x}^2 = \sigma^2_x(z_n) = \int_{-\infty}^{\infty} (s - \mu^n_x)^2\, p^n_x\, ds.$$ (4.2.7)

The minimal observed value will be denoted as

$$y_{0n} = \min_{1 \le i \le n} y_i.$$ (4.2.8)

The minimum of $f(x)$

$$s_0 = \min_{x \in A} f(x)$$ (4.2.9)

The conditional distribution of $f(x)$

$$F^n_x(s) = \int_{-\infty}^{s} p^n_x\, ds$$ (4.2.10)

and the limit

$$F^0_x = F^0_x(s) = \lim_{n \to \infty} F^s_x(s).$$ (4.2.11)

It is supposed that (4.2.11) holds at the continuity points of $F_x^0(s)$.
Denote the real function of the parameters a, b, c by

$$\phi = \phi_n = \phi(a, b, c), \tag{4.2.12}$$

where

$$a = a_x = a^n = d_x^n \in R \tag{4.2.13}$$

$$b = b_x = b^n = b_x^n \geq 0 \tag{4.2.14}$$

$$c = c_n \in R. \tag{4.2.15}$$

The upper and the lower indices n, x are present in some expressions of ϕ, a, b, c to emphasise that the function ϕ and the parameters a, b, c can depend on n and x. These indices will sometimes be omitted to make the expressions less complicated. The point of the minimum of ϕ_n will be denoted as

$$x_{n+1} \in \arg \min_{x \in A} \phi(d_x^n, b_x^n, c_n). \tag{4.2.16}$$

The distance from any fixed $x \in A$ to the nearest observation will be denoted as

$$r = r_n = r_n(x) = \min_{1 \leq i \leq n} \|x_i - x\| = \|x_{i(x)} - x\|. \tag{4.2.17}$$

Here $i(x)$ is the index of the nearest observation.
Set A will be divided into two parts B and C by the following conditions

$$B = \left\{ x : \lim_{n \to \infty} r_n(x) = 0 \right\} \tag{4.2.18}$$

$$C_\varepsilon = \left\{ x : \lim_{n \to \infty} r_n(x) = \varepsilon > 0 \right\} \tag{4.2.19}$$

and

$$C = \bigcup_{\varepsilon > 0} C_\varepsilon \tag{4.2.20}$$

The condition (4.2.18) means that if $n \to \infty$ then at least one observation will be in the neighbourhood of x. The conditions (4.2.19) and (4.2.20) mean that there exists such a neighbourhood of x which does not contain any observation independently of how many of them are made.

We shall consider a case without noise

$$g(x) = 0 \tag{4.2.21}$$

and a case with noise

$$g(x) \neq 0, \ E\{g(x)\} = 0, \ \text{var}\{g(x)\} \leq g < \infty. \tag{4.2.22}$$

Since the same assumptions and notations will be considered in different lemmas and theorems it seems more convenient to set them out and to number all of them from the start and to use the reference numbers in what follows.

4.2.23 Assume f is a continuous function of $x \in A$ and a measurable function of ω.

4.2.24 Assume set $A \in R^n$ is closed and bounded.

4.2.25 Denote $a_x^n = \mu_x^n$, where μ_x^n is from (4.2.6).

4.2.25^a Assume μ_x^n is a continuous function of x.

4.2.26 Denote $b_x^n = \sigma_x^n$, where σ_x^n is from (4.2.7).

4.2.26^a Denote $c_n = \min_{x \in A} \mu_x^n$.

4.2.26^b Assume $\min_{x \in A} \mu_x^n = y_{0n}$ where y_{0n} is from (4.2.8).

4.2.26^c Assume σ_x^n is a continuous function of x.

4.2.27 Assume there exists a limit $\lim_{n \to \infty} a_x^n = a_x^0 \in R$.

4.2.28 Assume there exists a limit $\lim_{n \to \infty} b_x^n = b_x^0 \in R$.

4.2.29 Assume there exists a limit $\lim_{n \to \infty} c_n = c_0 \in R$.

4.2.29^a Assume there exists a limit $\lim_{n \to \infty} F_x^n(s) = F_x^0(s)$ in the sense of (4.2.11) where $F_x^0(s)$ is a distribution function.

4.2.30   Assume $b_x^0 = 0$, if $x \in B$, where B is defined by (4.2.18).

4.2.31   Assume $b_x^0 > 0$, if $x \in C$, where C is defined by (4.2.20).

4.2.32   Assume there exists a conditional probability density p_x, $p_x > 0$, if $b_x > 0$.

4.2.33   Assume $F_x^n(s)$ is a continuous function of μ_x^n, σ_x^n at the continuity points of $F_x^0(s)$.

4.2.34   Assume $\sup_n |\min(c_n, s)|$ is uniformly integrable with respect to

$F_x^n(s)$.

4.2.35   Denote $\phi(a_x, b_x, c) = \int_{-\infty}^{\infty} \min(s, c)\, dF_x(s)$.

4.2.36   Assume a limit of μ_x^n when $n \to \infty$ is $\mu_x^0 = f(x)$ uniformly on $x \in B$.

4.2.36[a]   Assume $y_i = f(x_i)$ (this means zero noise).

4.2.37   Assume $\phi(a', b', c) > \phi(a'', b'', c)$ if $b'' > 0$, $b' = 0$.

4.2.37[a]   Assume that $f(s)$ is a continuous function of a, b, c.

4.2.37[b]   Assume that ϕ is a lower semicontinuous function of $x \in A$.

LEMMA 4.2.1.   *Assume conditions* (4.2.27) *to* (4.2.31), (4.2.37) *and* (4.2.37[a]) *hold (with the exception of* (4.2.29[a])*). Then*

$$\lim_{n \to \infty} \phi(a_{x''}^n, b_{x''}^n, c_n) < \lim_{n \to \infty} \phi(a_x^n, b_{x'}^n, c_n) \tag{4.2.38}$$

where $x'' \in C$, $x' \in B$.

Proof. From the continuity of ϕ (conditions (4.2.37[a])) and conditions (4.2.27) to (4.2.29) it follows that condition (4.2.38) is equivalent to the condition

$$\phi(a_{x'}^0, b_{x'}^0, c_0) > \phi(a_{x''}^0, b_{x''}^0, c_0) \tag{4.2.39}$$

The inequality (4.2.39) follows directly from conditions (4.2.27) to (4.2.31) and (4.2.37) (with the exception of (4.2.29^a)).

LEMMA 4.2.2. *Suppose the conditions of Lemma 4.2.1 hold and there exists*

$$x^* \in \arg\min_{x \in A} \phi.$$ (4.2.39^a)

Then

$$C = \emptyset.$$

Proof. Assume that there exists $x'' \in C$. Then from (4.2.31) it follows that $b^0_{x''} > 0$. From (4.2.30) it follows that $b^0_x = 0$ for all $x \in B$. The meaning of (4.2.38) is that there exists $\varepsilon > 0$ and n_ε such that

$$\phi(d^n_{x''}, b^n_{x''}, c_n) \leq \phi(a^n_{x'}, b^n_{x'}, c_n) - \varepsilon$$ (4.2.40)

if $n > n_\varepsilon$, $x' \in B$, $x'' \in C$.

Hence from the definitions (4.2.16) and the assumption (4.2.39^a) there follows the existence of n such that

$$x_{n+1} \bar{\in} B.$$

This means that

$$x_{n+1} \in A \backslash B = C$$

which contradicts the definition (4.2.20) of set C.

LEMMA 4.2.3. *Suppose that the conditions (4.2.25), (4.2.26), (4.2.26^a) and (4.2.27) to (4.2.32) hold. Then the function defined by (4.2.35) corresponds to the condition (4.2.37).*

Proof. From Chebyshev's inequality and conditions (4.2.25), (4.2.26), (4.2.27) and (4.2.28) it follows that for $k > 0$

$$\int_{|s - a_x| \geq k b_x} dF_x(s) \leq k^{1/2}$$

or

$$1 - F_x(a_x + kb_x) + F_x(a_x - kb_x) \leq k^{1/2} \tag{4.2.41}$$

Since F_x is a distribution function

$$0 \leq F_x(s) \leq 1 \tag{4.2.42}$$

and

$$F_x(s + \varepsilon) \geq F_x(s), \quad \varepsilon > 0. \tag{4.2.43}$$

In the case where $b_x^n \to 0$ and $a_x^n \to a_x^0$ from conditions (4.2.29^a) and (4.2.41) to (4.2.43) it follows that

$$F_x^0(s) = \begin{cases} 0, & \text{if } s < a_x^0 \\ 1, & \text{if } s \geq a_x^0 \end{cases} \tag{4.2.44}$$

In the case where $x \in B$ from conditions (4.2.29), (4.2.30), (4.2.35) and (4.2.44) it follows that

$$\phi(a_x^0, 0, c_0) = \min(a_x^0, c_0). \tag{4.2.45}$$

Hence in accordance with (4.2.26^a) it follows that

$$\phi(a_x^0, 0, c_0) = c_0, \quad \text{if } x \in B. \tag{4.2.46}$$

If $x \in C$, then in accordance with (4.2.31) the limit value

$$b_x^0 > 0. \tag{4.2.47}$$

Using assumption (4.2.32) the function (4.2.35) can be expressed as

$$\phi(a_x^0, b_x^0, c_0) = c - \int_{-\infty}^{c} (c - s) p_x \, ds. \tag{4.2.48}$$

Within the bounds of integration the difference $c - s > 0$ and, in accordance with assumption (4.2.32), the probability density $p_x > 0$, hence

$$\phi(a_x^0, b_x^0, c_0) < c_0, \quad \text{if } x \in C. \tag{4.2.49}$$

Under the assumptions (4.2.30) and (4.2.31) from the equalities (4.2.46) and (4.2.49) it follows that the function defined by (4.2.35) corresponds to the condition (4.2.37).

Define the real function $l_n(s)$, here $n = 1, 2, \ldots$ and $s \in R$.
Let

$$l(s) = \sup_n |l_n(s)| \qquad\qquad (4.2.52)$$

and

$$l_0(s) = \lim_{n \to \infty} l_n(s).$$

Suppose that

$l(s)$ is uniformly integrable with respect to $F^n(s)$. $\qquad (4.2.53)$

Assume that

$$l_n(s) \to l_0(s) \qquad\qquad (4.2.54)$$

where $l_n(s)$ is continuous in the interval $[-d, d]$.

LEMMA 4.2.4. *Suppose conditions* (4.2.29^a), (4.2.53) *and* (4.2.54) *hold. Then*

$$\lim_{n \to \infty} \int_{-\infty}^{\infty} l_n(s)\, dF^n(s) = \int_{-\infty}^{\infty} l_0(s)\, dF^0(s). \qquad (4.2.55)$$

Proof. It is obvious that

$$\left| \int_{-\infty}^{\infty} l_0(s)\, dF^0(s) - \int_{-\infty}^{\infty} l_0(s)\, dF^n(s) \right| \le \delta'_n + \delta''_n \qquad (4.2.56)$$

where

$$\delta'_n = \left| \int_{-\infty}^{\infty} l_0(s)\, dF^0(s) - \int_{-\infty}^{\infty} l_0(s)\, dF^n(s) \right|$$

and

$$\delta''_n = \left| \int_{-\infty}^{\infty} l_0(s)\, dF^n(s) - \int_{-\infty}^{\infty} l_n(s)\, dF^n(s) \right|$$

From conditions (4.2.53), (4.2.54) and the convergence theorem of Loev (1960) it follows that

$$\delta'_n < \varepsilon, \text{ if } n > n'_\varepsilon . \tag{4.2.57}$$

From the condition (4.2.54) it follows that

$$|l_0(s) - l_n(s)| < \varepsilon \text{ for all } s \in [-d, d] \tag{4.2.58}$$

if $n > n''_\varepsilon$.

From condition (4.2.53) and the convergence theorem of Loev (1960) it follows that

$$\int_{|s|>d} l_n(s) \, dF^n(s) < \varepsilon, \text{ if } n > n'''_\varepsilon \text{ and } d \geq d_\varepsilon, \text{ where } d_\varepsilon \to \infty \text{ when } \varepsilon \to 0. \tag{4.2.59}$$

From the condition (4.2.58)

$$\int_{-d}^{d} |l_0(s) - l_n(s)| \, dF^n(s) \leq \int_{-d}^{d} \varepsilon dF^n(y) \leq \varepsilon, \text{ if } n > n''_\varepsilon. \tag{4.2.60}$$

From conditions (4.2.57), (4.2.59) and (4.2.60) it follows that

$$\delta'_n + \delta''_n \leq 3\varepsilon \text{ if } n > n_\varepsilon \tag{4.2.61}$$

where

$$n_\varepsilon = \max (n'_\varepsilon, n''_\varepsilon, n'''_\varepsilon).$$

Condition (4.2.55) follows from conditions (4.2.56) and (4.2.61).

LEMMA 4.2.5. *Suppose the conditions (4.2.33) and (4.2.34) hold. Then ϕ defined by (4.2.35) is a continuous function of a, b, c.*

Proof. Let

$$l_n(s) = \min (s, c_n). \tag{4.2.62}$$

Hence in accordance with (4.2.35) we have

$$\phi(a, b, c) = \int_{-\infty}^{\infty} l_n(s) \, dF_x^n(s).$$

From condition (4.2.33) it follows that

$$\lim_{n \to \infty} F_x^n(s) = \lim_{\substack{\mu_{x_n}^n \to \mu_x^0 \\ \sigma_x^n \to \sigma_x^0}} F_x^n(s) = F_x^0(s) \tag{4.2.63}$$

where the equality (4.2.63) holds at the continuity points of $F_x^0(s)$.

From conditions (4.2.52) and (4.2.62) it follows that

$$l(s) = \sup_n |l_n(s)| = \sup_n |\min(s, c_n)| \tag{4.2.64}$$

Hence (4.2.53) follows from (4.2.34).

It is obvious that $l_n(s)$ defined by (4.2.62) is a continuous function. Since condition (4.2.54) holds (4.2.29^a) follows directly from (4.2.63).

So all the conditions of Lemma 4.2.4 hold. This means that the expression (4.2.55) is true and ϕ is a continuous function.

LEMMA 4.2.6. *Suppose that conditions* (4.2.24), (4.2.25), (4.2.26), (4.2.26^a), (4.2.26^b) *and* (4.2.27) *to* (4.2.35) *hold. Assume that either* (25^a), (26^c) *or* (37^b) *is true. Then*

$$C = \varnothing. \tag{4.2.65}$$

Proof. In accordance with Lemma 4.2.2 the condition (4.2.65) is true if the conditions (4.2.27) to (4.2.31) hold and function ϕ corresponds to (4.2.37), (4.2.37^a) and 4.2.39^a). In accordance with Lemma 4.2.5 the condition (4.2.37^a) follows from   (4.2.33) to (4.2.35). Condition (4.2.37) follows from Lemma 4.2.3 and conditions (4.2.25), (4.2.26), (4.2.26^a) and (4.2.27) to (4.2.32). The existence of $\min_{x \in A} \phi$ (condition (4.2.39^a) follows from (4.2.24), (4.2.25^a), (4.2.26^c) and (4.2.37^a), otherwise it follows from (4.2.24) and (4.2.37^b).

LEMMA 4.2.7. *Suppose condition* (4.2.36^a) *holds and the point of final decision is*

$$x_{0n} \in \arg \min_{x \in A} \mu_x^n. \tag{4.2.66}$$

Then under the conditions of Lemma 4.2.6

$$\lim_{n \to \infty} c_n = s_0 \tag{4.2.68}$$

for all continuous functions $f(x)$ where s_0 is the minimum of $f(x)$.

Proof. From conditions $(4.2.25)$, $(4.2.26^b)$ and $(4.2.36^a)$ it follows that

$$\arg \min_{x \in A} a_x^n = \arg \min_{1 \le i \le n} f(x_i).$$

Hence expression $(4.2.66)$ can be represented as

$$x_{0n} \in \arg \min_{1 \le i \le n} f(x_i) \tag{4.2.69}$$

From the definition $(4.2.20)$ of C and Lemma 4.2.6 it follows that the distance to the nearest observation $r_n(x) \to 0$ when $n \to \infty$ for any $x \in A$.

Hence $(4.2.68)$ is true for any continuous function because $(4.2.26^a)$ and $(4.2.69)$ hold.

LEMMA 4.2.8. *Suppose the assumptions of Lemma 4.2.6 and conditions $(4.2.36)$ hold. Then*

$$\lim_{n \to \infty} c_n = s_0 = \min_{x \in A} f(x) \tag{4.2.70}$$

for any continuous function $f(x)$.

Proof. From $C = \emptyset$ (Lemma 4.2.6) and from definitions $(4.2.18)$ to $(4.2.20)$ it follows that

$$B = A.$$

From this equality and assumption $(4.2.36)$ it follows that

$$\lim_{n \to \infty} \sup_{x \in A} |\mu_x^n - f(x)| = 0. \tag{4.2.71}$$

Hence for any $\varepsilon > 0$ there exists an integer n_ε such that

$$\sup_{x \in A} |\mu_x^n - f(x)| < \varepsilon \text{ if } n > n_\varepsilon. \tag{4.2.72}$$

Hence

$$| \inf_{x \in A} \mu_x^n - \min_{x \in A} f(x)| < \varepsilon \text{ if } n > n_\varepsilon . \tag{4.2.73}$$

From (4.2.73), condition (4.2.70) follows directly.
Under conditions (4.2.26^a), (4.2.26^b) and 4.2.36^a)

$$c_n = \min_{x \in A} \mu_x^n = \min_{1 \le i \le n} f(x_i) = y_{0n}$$

and

$$c_{n+1} = \min (y_{0n}, y) \tag{4.2.74}$$

where y_{0n} is minimal from n observations and y is the result of the $(n + 1)$-th observation. Then from (2.5.1) and (2.5.5) the one-step Bayesian method is

$$x_{n+1} \in \arg\min_{x \in A} E\{c_{n+1}/z_n\} \tag{4.2.75}$$

where $E\{c_{n+1}/z_n\}$ is the conditional expectation of c_{n+1} with regard to z_n. From (4.2.24) and (4.2.36^a) it follows that

$$x_{n+1} \in \arg\min_{x \in A} \int_{-\infty}^{\infty} \min (y_{0n}, y) \, p_x^n \, dy \tag{4.2.76}$$

Now let us consider the case without noise when $g = 0$ and (4.2.36^a) holds.

THEOREM 4.2.9. *Suppose that conditions of Lemma* 4.2.6. *and the assumption* (4.2.36^a) *hold. Then the Bayesian method* (4.2.75) *converges to the minimum of any continuous function in the sense that*

$$\lim_{n \to \infty} f(x_{0n}) = s_0 . \tag{4.2.77}$$

Proof. Theorem 4.2.9 follows from Lemma 4.2.7 since the expression (4.2.76) is a special case of (4.2.35).

When the error is present ($g \ne 0$) then, strictly speaking, the formula (4.2.76) cannot be regarded as the Bayesian method (4.2.75) because $c_{n+1} = \min (y_{0n}, y)$

only when $c_n = y_{0n}$ and $f(x_{n+1}) = y$, and that is not true in the case with noise. In this case it would be more natural to substitute c_n for y_{0n} in the expression (4.2.76) and to define the one-step Bayesian method as

$$x_{n+1} \in \arg\min_{x \in A} \int_{-\infty}^{\infty} \min (c_n, s) \, p_x^n \, ds. \tag{4.2.79}$$

In the presence of noise we have not only the *a priori* probability P which is defined on the space of continuous functions $f(x, \omega)$ to be minimised, but also the probability P' defined on the space of noise functions $g(x, \omega)$. In such a case condition (4.2.36) could be reasonably expected to hold only with some probability P'.

THEOREM 4.2.10. *Suppose that the conditions of Lemma 4.2.6 hold. Assume that condition* (4.2.36) *holds with probability P'. Then the Bayesian method* (4.2.79) *converges to the minimum of any continuous function in the sense of* (4.2.77) *with probability P'*

Proof. Theorem 4.2.10 follows from Lemma 4.2.8 because the expression (4.2.79) is a special case of (4.2.35). It is obvious from the proof of Lemma 4.2.8 that if (4.2.36) holds with probability P' then (4.2.70) also holds with probability P'.

EXAMPLE 4.2.1. Suppose $f(x)$ is a Wiener process. Let us test sufficient convergent conditions. A positive answer is given by the following propositions.

PROPOSITION 4.2.11. *(Conditions* (4.2.27) *and* (4.2.36)). *There exists a limit of conditional expectation of the Wiener process, when $n \to \infty$ for any $x \in A$ and*

$$\lim \mu_x^n = f(x) \tag{4.2.80}$$

uniformly on B.

Proof. Let

$$\varepsilon_1 = \lim_{n \to \infty} \Delta_1 \quad \text{and} \quad \varepsilon_2 = \lim_{n \to \infty} \Delta_2$$

where $\Delta_1 = |x_i - x|$, $\Delta_2 = |x - x_{i+1}|$ and x_i, x_{i-1} are neighbours of x.

Consider the following four cases:

1) $\varepsilon_1 = 0$ and $\varepsilon_2 = 0$,

2) $\varepsilon_1 = 0$, $\varepsilon_2 > 0$,

3) $\varepsilon_1 > 0$, $\varepsilon_2 = 0$,

4) $\varepsilon_1 > 0$, $\varepsilon_2 > 0$, here $x \in C$.

The conditional expectation of the Wiener process is

$$\mu_x^n = \frac{f(x + \Delta_2)\Delta_1 + f(x - \Delta_1)\Delta_2}{\Delta_1 + \Delta_2}.$$

In the first case

$$\lim_{n \to \infty} \mu_x^n = \lim_{\substack{\Delta_1 \to 0 \\ \Delta_2 \to 0}} \mu_x^n = f(x). \qquad (4.2.81)$$

In the second case

$$\lim_{n \to \infty} \mu_x^n = \lim_{\substack{\Delta_1 \to 0 \\ \Delta_2 \to \varepsilon_2}} \mu_x^n = f(x) \qquad (4.2.82)$$

because

$$\lim_{\Delta_1 \to 0} \mu_x^n = f(x)$$

and

$$\lim_{\Delta_2 \to \varepsilon_2} \mu_x^n = \frac{f(x + \varepsilon_2)\Delta_1 + f(x - \Delta_1)\varepsilon_2}{\Delta_1 + \varepsilon_2}.$$

In the third case

$$\lim_{n \to \infty} \mu_x^n = \lim_{\substack{\Delta_1 \to \varepsilon_1 \\ \Delta_2 \to 0}} \mu_x^n = f(x). \qquad (4.2.83)$$

In the fourth case

$$\lim_{n \to \infty} \mu_x^n = \lim_{\substack{\Delta_1 \to \varepsilon_1 \\ \Delta_2 \to \varepsilon_2}} \mu_x^n = \frac{f(x + \varepsilon_2)\varepsilon_1 + f(x - \varepsilon_1)\varepsilon_2}{\varepsilon_1 + \varepsilon_2}. \tag{4.2.84}$$

In all cases the convergence is uniform.

PROPOSITION 4.2.12. (*Conditions* (4.2.28), (4.2.30) *and* (4.2.31)). *There exists a limit of conditional variance of the Wiener process when* $n \to \infty$ *and*

$$\lim_{n \to \infty} \sigma_x^n = 0, \ \text{if } x \in B \tag{4.2.85}$$

and

$$\lim_{n \to \infty} \sigma_x^n > 0, \ \text{if } x \in C \tag{4.2.86}$$

Proof.. The conditional variance of the Wiener process can be expressed by

$$\sigma_x^{n^2} = \frac{\Delta_1 \Delta_2}{\Delta_1 + \Delta_2} \tag{4.2.87}$$

We shall consider the same four cases as in the previous proposition.
In the first case

$$\lim_{n \to \infty} \sigma_x^n = \lim_{\substack{\Delta_1 \to 0 \\ \Delta_2 \to 0}} \sigma_x^n = 0 \tag{4.2.88}$$

because there exist limits

$$\lim_{\Delta_1 \to 0} \frac{\Delta_1 \Delta_2}{\Delta_1 + \Delta_2} = 0$$

and

$$\lim_{\Delta_2 \to 0} \frac{\Delta_1 \Delta_2}{\Delta_1 + \Delta_2} = 0 \ .$$

It is easy to see that in the second and third cases

$$\lim_{n \to \infty} \sigma_x^n = 0 \tag{4.2.89}$$

In the fourth case

$$\lim_{n \to \infty} \sigma_x^n = \lim_{\substack{\Delta_1 \to \varepsilon_1 \\ \Delta_2 \to \varepsilon_2}} \sigma_x^n = \frac{\varepsilon_1 \varepsilon_2}{\varepsilon_1 + \varepsilon_2} > 0, \qquad (4.2.90)$$

because

$$\lim_{\Delta_1 \to \varepsilon_1} \frac{\Delta_1 \Delta_2}{\Delta_1 + \Delta_2} = \frac{\varepsilon_1 \Delta_2}{\varepsilon_1 + \Delta_2} \quad ,$$

$$\lim_{\Delta_2 \to \varepsilon_2} \frac{\Delta_1 \Delta_2}{\Delta_1 + \Delta_2} = \frac{\varepsilon_2 \Delta_1}{\Delta_1 + \varepsilon_2} \quad ,$$

and

$$\lim_{\Delta_1 \to \varepsilon_1} \left(\lim_{\Delta_2 \to \varepsilon_2} \frac{\Delta_1 \Delta_2}{\Delta_1 + \Delta_2} \right) = \lim_{\Delta_1 \to \varepsilon_1} \frac{\varepsilon_2 \Delta_1}{\Delta_1 + \varepsilon_2} = \frac{\varepsilon_1 \varepsilon_2}{\varepsilon_1 + \varepsilon_2} \quad .$$

PROPOSITION 4.2.13. (*Condition 4.2.29*). *There exists a limit c_0 of a minimum c_n of conditional expectation μ_x^n when $n \to \infty$.*

Proof. Since the conditional expectation μ_x^n of the Wiener process is a piecewise function, then

$$c_n = \min_{x \in A} \mu_x^n = y_{0n} \geq s_0 = \min_{x \in A} f(x). \qquad (4.2.91)$$

Here $y_{0n} = \min_{1 \leq i \leq n} y_i$ is a non-increasing function. Hence there exists such $c_0 > s$ that

$$\lim_{n \to \infty} y_{0n} = 0. \qquad (4.2.92)$$

PROPOSITION 4.2.14. (*Conditions (4.2.29ª) and (4.2.33)*). *The distribution function $F_x^n(s)$ of a Wiener process is continuous with regard to conditional expectation μ_x^n and conditional variation $\sigma_x^{n^2}$ at points where $F_x^n(s)$ is continuous and there exists a limit distribution function $F_x^0(s)$.*

Proof. There exists a limit

$$\lim_{\substack{\mu_x^n \to a_x^0 \\ \sigma_x^n \to b_x^0}} F_x^n(s) = F_x^0(s) \tag{4.2.93}$$

at the continuity points of $F_x^0(s)$.

This is because there exists the limit

$$\lim_{\sigma_x^n \to b_x^0} F_x^n(s) = \begin{cases} (\sqrt{(2\pi)}\, b_x^0)^{-1} \int_{-\infty}^{s} \exp\left(-1/2\left(\dfrac{s-\mu_x^n}{b_x^0}\right)^2\right) ds, & b_x^0 > 0 \\ 0, & b_x^0 = 0, \ s < \mu_x^n \\ 1, & b_x^0 = 1, \ s > \mu_x^n \end{cases} \tag{4.2.94}$$

uniformly on $0 < |\mu_x^n - a_x^0| < \delta = a_x^0 - s$

and the limit

$$\lim_{\mu_x^n \to a_x^0} F_x^n(s) = \begin{cases} (\sqrt{(2\pi)}\, \sigma_x^n)^{-1} \int_{-\infty}^{s} \exp\left(-1/2\left(\dfrac{s-a_x^0}{\sigma_x^n}\right)^2\right) ds, & \sigma_x^n > 0 \\ 0, & \sigma_x^n = 0, \ s < a_x^n \\ 1, & \sigma_x^n = 0, \ s > a_x^n \end{cases} \tag{4.2.95}$$

at any $0 < |\sigma_x^n - b_x^n| < \delta$.

Equality (4.2.94) is true, because there exists the limit

$$\lim_{\sigma_x^n \to b_x^0} \sup_{\mu_x^n \in B_s} F_x^n(s) \tag{4.2.96}$$

where

$$B_s = \{\mu_x^n : 0 < |\mu_x^n - a_x^0| < \delta\}$$

and

$$\delta < |a_x^0 - s|.$$

It is easy to see that

$$\sup_{\mu_x^n \in B_s} (\sqrt{2\pi}\, \sigma_x^n)^{-1} \int_{-\infty}^{s} \exp\left(-1/2\left(\frac{\tau - \mu_x^n}{\sigma_x^n}\right)^2\right) d\tau$$

$$= (\sqrt{2\pi}\, \sigma_x^n)^{-1} \int_{-\infty}^{s} \exp\left(-1/2\left(\frac{\tau - \mu_x'}{\sigma_x^n}\right)^2\right) d\tau. \tag{4.2.97}$$

Here $\mu_x' = s + \varepsilon$ if $a_x^0 > s$

and $\mu_x' = a - s + \varepsilon$ if $a_x^0 < s$.

From (4.2.96) and (4.2.97) it follows that

$$\lim_{\sigma_x^n \to b_x^0} \sup_{\mu_x^n \in B_s} F_x^n(s) = \begin{cases} (\sqrt{2\pi}\, b_x^0)^{-1} \int_{-\infty}^{s} \exp\left(-1/2\left(\frac{\tau - \mu_x'}{b_x^0}\right)^2\right) d\tau, & b_x^0 > 0 \\ 0, & b_x^0 = 0,\, s < a_x^n \\ 1, & b_x^0 = 0,\, s > a_x^n \end{cases}$$

$$\tag{4.2.98}$$

The function $F_x^n(s)$ does not depend directly on n when μ_x^n and σ_x^n are fixed. In such a case the existence of the limit

$$\lim_{n \to \infty} F_x^n(s) = F_x^0(s)$$

follows from the existence of limit (4.2.96) and the existence of limits (4.2.80) and (4.2.85).

PROPOSITION 4.2.15. (*Condition* (4.2.34)). *Function* $|s|$ *is uniformly integrable with respect to* $F_x^n(s)$.

Proof. Let

$$\psi_n'(c) = \sup_{\substack{0 \le \sigma_x^n \le b \\ a_1 \le \mu_x^n \le a_2}} \int_c^{\infty} |s|\, dF_x^n(s).$$

$$\tag{4.2.99}$$

and

$$\psi_n''(c) = \sup_{\substack{0 \le \sigma_x^n \le b \\ a_1 \le \mu_x^n \le a_2}} \int_{-\infty}^{-c} |s| \, dF_x^n(s).$$

(4.2.100)

Here

$$\int_c^\infty |s|\,dF_x^n(s) = \begin{cases} (\sqrt{(2\pi)}\,\sigma_x^n)^{-1} \int_c^\infty |s| \exp\left(-1/2\left(\frac{s-\mu_x^n}{\sigma^n}\right)^2\right) ds, & \sigma_x^n > 0 \\ 0, & \sigma_x^n = 0,\ c > \mu_x^n \\ 1, & \sigma_x^n = 0,\ c < \mu_x^n \end{cases}$$

and

$$\int_{-\infty}^{-c} |s|\,dF_x^n(s) = \begin{cases} (\sqrt{(2\pi)}\,\sigma_x^n)^{-1} \int_{-\infty}^{-c} |s| \exp\left(-1/2\left(\frac{s-\mu_x^n}{\sigma^n}\right)^2\right) ds, & \sigma_x^n > 0 \\ 0, & \sigma_x^n = 0,\ -c < \mu_x^n \\ 1, & \sigma_x^n = 0,\ -c > \mu_x^n \end{cases}$$

where

$$b = \max_{x \in A} \sigma_x^n, \quad a_1 = \min_{x \in A} \mu_x^n, \quad a_2 = \max_{x \in A} \mu_x^n$$

When $|c| > a_2$ then the upper limit (4.2.99) is reached at $\sigma_x^n = b$ and $\mu_x^n = a_2$ because the integral (4.2.99) is nondecreasing when μ_x^n and σ_x^n increase.

The upper limit of (4.2.100) is at the point $\sigma_x^n = b$ and $\mu_x^n = a_1$ if $-c < a_1$. Hence

$$\psi_n'(c) = \begin{cases} (\sqrt{(2\pi)}\,b)^{-1} \int_c^\infty |s| \exp\left(-1/2\left(\frac{s-a_2}{b}\right)^2\right) ds, & b > 0 \\ 0, & b = 0,\ c > a_2 \\ 1, & b = 0,\ c < a_2 \end{cases}$$

(4.2.101)

and

$$\psi_n''(c) = \begin{cases} (\sqrt{(2\pi b\!\!\!/}\,)^{-1} \int_{-\infty}^{-c} |s| \exp{(-1/2(\frac{s-a_1}{b})^2)}\,ds & , \ b > 0 \\ 0, & b = 0, -c < a_1 \\ 1, & b = 0, -c > a_1 . \end{cases}$$

$$(4.2.102)$$

It follows from (4.2.101) and (4.2.102) that for any $\varepsilon > 0$ there exists c_ε such that

$$\psi_n'(c) < \varepsilon, \ c > c_\varepsilon,$$

and

$$\psi_n''(c) < \varepsilon, \ c > c_\varepsilon.$$

Hence

$$\psi_n(c) = \psi_n'(c) + \psi_n''(c) < 2\varepsilon, \ c > c_\varepsilon, \qquad (4.2.103)$$

independently of n where

$$\psi_n(c) = \int_{|s|>c} |s|\, dF_x^n(s).$$

Since $\psi_n(c) \to 0$ uniformly with respect to n, this means (see Loev (1960) that $|s|$ is uniformly integrable with respect to $F_x^n(s)$.

Since

$$\sup_n |\min{(c_n, s)}| \le |s|$$

it follows that condition (4.2.34) is true.

PROPOSITION 4.2.16. (*Conditions* (4.2.37), (4.2.37^a) *and* 4.2.37^b)). *Function*

$$\phi(a, b, c) = (\sqrt{(2\pi b\!\!\!/}\,)^{-1} \int_{-\infty}^{\infty} \min{(s, c)} \exp{(-1/2(\frac{s-a}{b})^2)}\,ds \qquad (4.2.104)$$

where

$$0 = \mu_x^n, \; b = \sigma_x^n \text{ and } c = y_{0n}$$

is a continuous function of a, b, c, x and

$$\phi(a', \; b', c) \; < \; \phi(a'', b'', c) \tag{4.2.105}$$

if b" > 0 and b' = 0.

Proof. The continuity of function ϕ with respect to a, b, c can be proved in the same way as the continuity of function $F_x^n(s)$ with respect to μ_x^n, σ_x^n in Proposition 4.2.14. The limit

$$\lim_{b \to 0} \; (\sqrt{(2\pi}b\,)^{-1} \int_{-\infty}^{s} \exp\left(-1/2\left(\frac{s-a}{b}\right)^2\right) ds \; = \; \begin{cases} 0, \; s < a \\ 1, \; s > a \end{cases} \tag{4.2.106}$$

From (4.2.104) and (4.2.106)

$$\phi(a', \; b', c) \; = \; \min(a, c) \; = \; c \tag{4.2.107}$$

because here

$$c \; = \; \min_{x \in A} \mu_x^n \tag{4.2.108}$$

and

$$a \; = \; \mu_x^n. \tag{4.2.109}$$

Formula (4.2.104) can be expressed in the following way

$$\phi(a, b, c) \; = \; c - (\sqrt{(2\pi)}b\,)^{-1} \int_{-\infty}^{c} (c - s) \exp\left(-1/2\left(\frac{s-a}{b}\right)^2\right) ds$$

The second part of this expression is positive if b > 0 independently of |a| < ∞ and so

$$\phi(a'', \; b'', c) \; < \; \phi(a', \; b', c)$$

if b" > 0 and b' = 0.

Thus conditions (4.2.27) to (4.2.31), (4.2.33), (4.2.34), (4.2.36), (4.2.37) to (4.2.37[b]) are proved. The remaining conditions (4.2.25[a]), (4.2.26[b]), (4.2.26[c]), (4.2.32) and (4.2.36[a]) are obvious in the case of the Wiener process.

Similar results were obtained by Senkiene (1980) for the Wiener process with Gaussian noise.

It is clear that any method of search which satisfies the condition of asymptotic density (4.2.65) converges to the global minimum of any continuous function $f(x)$ defined on a compact set A. The distinctive property of the Bayesian methods is that they are generally not uniform and provide greater density of observations in the areas around the best observed values of function f, under the following assumptions.

Suppose that there exists sets $S_i \subset A, i = 1, \dots, n$ such that

$$\bigcup_{i=1}^{n} S_i = A, \ S_i \cap S_j = \varnothing, \ \text{if } i \neq j \tag{4.2.110}$$

$$x_i \in S_i, \tag{4.2.111}$$

$$x_j \bar{\in} S_i, \ \text{if } j \neq i. \tag{4.2.112}$$

Assume that

$$\sigma_{x''}^n > \sigma_{x'}^n, \ \text{iff } r_{x''} > r_{x'},$$

$$\sigma_{x''}^n = \sigma_{x'}^n, \ \text{iff } r_{x''} = r_{x'} \tag{4.2.113}$$

independently of $i = 1, \dots, n$ for any $n = 1, 2, \dots$. Here

$$x' \text{ and } x'' \in S_i,$$

$$r_x = \|x - x_i\|, \ x \text{ and } x_i \in S_i.$$

Suppose that

$$\sup_{x \in S_0} \mu_x^n < \inf_{x \in S_i} \mu_x^n \tag{4.2.114}$$

if S_i has no common borders with S_0, where S_0 is a set S_i around the best observation point $x_0 = \arg \min_{1 \leq i \leq n} f(x_i)$.

Assume the risk function

$$\phi(\mu, \sigma, c) > \phi(\mu, \sigma + \beta, c),$$

$$\phi(\mu, \sigma, c) > \phi(\mu - \alpha, \sigma, c), \qquad\qquad (4.2.115)$$

where $\alpha > 0$, $\beta > 0$, and that ϕ is a continuous function of μ, σ, c.
Denote

$$r_i \;=\; \|x_i' - x_i\|$$

$$r_0 \;=\; \|x_0' - x_0\|$$

where

$$x_i' \in \arg\min_{x \in S_i} \phi(\mu, \sigma, c)$$

$$r_0' \in \arg\min_{x \in S_0} \phi(\mu, \sigma, c).$$

It is natural and convenient to define the relative density of observations as the relation K_i of radii r_i and r_0

$$K_i \;=\; r_i / r_0$$

or, in accordance with condition (4.2.113), as a relation K_i^s of standard deviations σ_i and σ_0 where $\sigma_i = \sigma_x^n$, $x = x_i'$ and $\sigma_0 = \sigma_x^n$, $x = x_0'$.

LEMMA 4.2.1. *If conditions (4.2.110) to (4.2.112), (4.2.114) and (4.2.115) hold then*

$$K_i^s \;=\; \sigma_i / \sigma_0 > 1 \qquad\qquad (4.2.116)$$

for such indices i where a corresponding set S_i has no common border with S_0.

Proof. From (4.2.115) and the continuity of function ϕ it follows that for any $\alpha > 0$ there exists $\beta > 0$ such that

$$\phi(\mu - \alpha, \sigma - \beta, c) \;=\; \phi(\mu, \sigma, c) \qquad\qquad (4.2.116.1)$$

From (4.2.116.1), (4.2.114) and (4.2.115) it follows that the minimum of function ϕ in the areas S_i will correspond to larger σ_x^n than in the area S_0 because in the area S_i the mean values μ_x^n are assumed to be larger, see (4.2.114).

THEOREM 4.2.12. *If the assumptions (4.2.110) to (4.2.115) are true then*

$$K_i = r_i/r_0 > 1. \qquad (4.2.117)$$

Here S_i has no common borders with S_0.

Proof. Inequality (4.2.117) follows from conditions (4.2.116) and (4.2.113).

Inequality (4.2.117) means that the distance to the next observation will be smaller in the area around the point of best observation x_0.

The condition (4.2.117) deterimines the ratio of radii r_i and r_0 corresponding to the sets S_i and S_0.

Now we shall define the value of the ratio K_i for the Gaussian distribution of $f(x)$ assuming that the values of the risk function ϕ are the same in both sets S_i and S_0.

This assumption is not restrictive when n is sufficiently large because it follows from (4.2.35), (4.2.30), (4.2.65), (4.2.68) and (4.2.70) that the risk function (4.2.79) will converge to the same limit S_0 for all sets S_i.

In this case the ratio of standard deviations

$$K_a^s = \sigma_a/\sigma_0 = \frac{\mu_a - y_{0n} + \varepsilon}{\mu_0 - y_{0n} + \varepsilon} \qquad (4.2.118)$$

and from (4.2.30) and (4.2.77) the asymptotic relation

$$K^s = \lim_{n \to \infty} K_a^s = \frac{f_a - f_0 + \varepsilon}{\varepsilon} \qquad (4.2.119)$$

Ratio (4.2.118) was determined from the equality of values of parameter $a = (c - \mu)/\sigma$ in the sets S_a and S_0. The parameter a defines the risk function in the Gaussian case, see expressions (5.1.1), (5.1.3), (5.1.5) and (5.1.6).

In the expressions (4.2.118) and (4.2.119) the following notation was used:

$$\mu_a = \mu_x^n, \text{ if } x = x_a'$$

$$\sigma_a = \sigma_x^n, \text{ if } x = x_a'$$

$$f_a = (L(A))^{-1} \int_A f(x) \, dx$$

$$f_0 = \min_{x \in A} f(x).$$

$L(A)$ is Lebesgue measure (the volume) of set A.

$$x_a' \in \arg \min_{x \in S_a} \phi(\mu, \sigma, c).$$

S_a is a set S_i around the observation where result y_i is nearest to the mean value f_a of function $f(x)$.

If the standard deviations are defined by expression (5.2.8) then it follows from (4.2.119) and (5.2.8) that

$$K = \lim_{n \to \infty} K_a = \lim_{n \to \infty} r_a / r_0 = \lim_{n \to \infty} \frac{(\sigma_a)^{1/2}}{(\sigma_0)^{1/2}} = \left(\frac{f_a - f_0 + \varepsilon}{\varepsilon} \right)^{1/2} \qquad (4.2.120)$$

In the case (5.2.7)

$$r_a = \|x_a' - x_a\| = \max_{x \in S_a} \|x - x_a\|$$

and

$$r_0 = \|x_0' - x_0\| = \max_{x \in S_0} \|x - x_0\|.$$

The relations (4.2.119) and (4.2.120) can be used to compare the efficiency of non-uniform Bayesian search with uniform search. In this case radius r_a roughly corresponds to the uniform search and radius r_0 to the Bayesian search around the best point and asymptotically around the global minimum. Thus the relation K can be regarded as the relative asymptotic density of observations of the Bayesian search round the global minimum compared with the uniform search.

It is clear from expressions (4.2.118) and (4.2.120) that the relative density is a decreasing function of the correction parameter ε. The relative density converges to 1 when $\varepsilon \to \infty$ and increases without bounds when $\varepsilon \to 0$. This means that if parameter ε is large then the search is nearly uniform. If the parameter ε is small then only an insignificant part of the observations will be made outside the region of the best observed point. This corresponds well to the meaning of the correction parameter ε as defined in sections 2.5 and 5.1.

4.3. The Gaussian field

If the probability densities (4.2.3) are Gaussian then the corresponding stochastic function is called the Gaussian field. We shall denote the vectors of expectation by μ^s, μ^v and μ and the matrices of covariance by Σ^s, Σ^v and Σ, respectively, where

$$\mu^s = (\mu_i^s), \ \mu^v = (\mu_i^v), \ \mu = (\mu_i), \ i = 1, \dots, l \tag{4.3.1}$$

$$\Sigma^s = (\sigma_{ij}^s), \ \Sigma^v = (\sigma_{ij}^v), \ \Sigma = (\sigma_{ij}), \ i, j = 1, \dots, l. \tag{4.3.2}$$

Here μ_i^s, μ_i^v, μ_i are expectations of the random variables $f(x_i)$, $g(x_i)$, $h(x_i)$ and σ_{ij}^s, σ_{ij}^v, σ_{ij} are covariances of the pairs $f(x_i)$, $f(x_j)$; $g(x_i)$, $g(x_j)$ and $h(x_i)$, $h(x_j)$, respectively. It is well known (see Anderson (1958)) that the conditional probability density will be Gaussian with expectation

$$\mu_x^n = \mu_x^s + \Sigma_x \Sigma^{-1}(y - \mu) \tag{4.3.3}$$

and variance

$$(\sigma_x^n)^2 = (\sigma_x^s)^2 - \Sigma_x \Sigma^{-1} \Sigma_x^T \tag{4.3.4}$$

where $y = (y_i)$, $y_i = h(x_i)$, $i = 1, \dots, n$. Here μ_x^s and $(\sigma_x^s)^2$ are the expectation and variance of the objective function $f(x)$, μ_x^n and $(\sigma_x^n)^2$ are the conditional expectation and variance of $f(x)$ when vector $z_n = (x_i, y_i, i = 1, 2, \dots, n)$ is fixed.

It is usually assumed that the expectation of noise is zero. Therefore the components σ_{xx_i}, $i = 1, \dots, n$ of the vector Σ_x should be the covariance of the pairs $f(x)$, $f(x_i)$.

In some cases the following expressions can be more convenient.
The conditional variance of $f(x)$ when z_n is fixed

$$\sigma_{xx}^n = \sigma_{xx}^{n-1} - \frac{\left(\sigma_{xx_n}^{n-1}\right)^2}{\sigma_{x_n x_n}^{*n-1}} \tag{4.3.5}$$

Here the conditional covariance of $f(x), f(x_n)$ when z_{n-1} is fixed

$$\sigma_{xx_n}^{n-1} = \sigma_{xx_n}^{n-2} - \frac{\sigma_{xx_{n-1}}^{n-2}\,\sigma_{x_nx_{n-1}}^{n-2}}{\sigma_{x_{n-1}x_{n-1}}^{*n-2}} \tag{4.3.6}$$

and the conditional variance of

$$h(x_n) = f(x_n) + g(x_n)$$

at fixed z_n is

$$\sigma_{x_nx_n}^{*n-1} = \sigma_{x_nx_n}^{*n-2} - \frac{\left(\sigma_{x_nx_{n-1}}^{*n-2}\right)^2}{\sigma_{x_{n-1}x_{n-1}}^{*n-2}} \ . \tag{4.3.7}$$

The conditional covariance of $h(x_n), h(x_{n-1})$ at fixed z_n is

$$\sigma_{x_nx_{n-1}}^{*n-1} = \sigma_{x_nx_{n-1}}^{*n-2} - \frac{\sigma_{x_nx_{n-2}}^{*n-2}\,\sigma_{x_{n-1}x_{n-2}}^{*n-2}}{\sigma_{x_{n-2}x_{n-2}}^{*n-2}} \tag{4.3.8}$$

The conditional expectation of $f(x)$ when z_n is fixed

$$\mu_x^n = \mu_x^{n-1} + \frac{\sigma_{xx_n}^{n-1}\left(y_n - \mu_{x_n}^{n-1}\right)}{\sigma_{x_nx_n}^{*n-1}} \tag{4.3.9}$$

The conditional expectation of $f(x_n)$ at fixed z_{n-1} is

$$\mu_{x_n}^{n-1} = \mu_{x_n}^{n-2} + \frac{\sigma_{x_nx_{n-1}}^{n-2}\left(y_{n-1} - \mu_{x_{n-1}}^{n-2}\right)}{\sigma_{x_{n-1}x_{n-1}}^{*n-2}} \ . \tag{4.3.10}$$

The zero indices correspond to unconditional parameters

$$\mu_x^0 = \mu_x^s$$

$$\sigma_{xx}^0 = (\sigma_x^s)^2$$

$$\sigma_{x_1 x_1}^{*0} = \sigma_{x_1}^2$$

$$\sigma_{x_2 x_1}^{*0} = \sigma_{x_2 x_1}$$

$$\sigma_{xx_1}^0 = \sigma_{xx_1}$$

EXAMPLE 4.3.1. In the case of independent noise with unit variance

$$\sigma_{x_n x_n}^{*0} = \sigma_{x_n x_n}^0 + 1$$

and

$$\sigma_{xx_n}^{*0} = \begin{cases} 0, & x \neq x_n \\ 1, & x = x_n \end{cases}$$

It follows from (4.3.7) that

$$\sigma_{x_n x_n}^{*n-1} = \sigma_{x_n x_n}^{n-1} + 1.$$

4.4 Homogeneous Wiener field

The only argument for considering the Gaussian field as a stochastic model so far was the fact that it is commonly used and well investigated.

It would be useful to derive some stochastic model more directly from the assumptions relevant to the problem of global optimization.

The first assumption is the continuity of samples. This means that probability P, to be a continuous function, should be 1.

The next natural assumption is homogeneity, which means that the l-dimensional distributions F defined in (4.2.2) should not depend on the origin of co-ordinates.

A third assumption is desirable to narrow the class of possible stochastic functions but here the choice is less obvious.

It is only clear that a stochastic model should not contradict the first two assumptions and *a priori* concepts about the objective function and be as simple as possible for the computation of conditional probabilites. The computation of conditional probabilities is simplest in cases when $f(x_i)$ and $f(x_j)$, $i \neq j$, are independent. However it contradicts the continuity assumption. It also contradicts *a priori* notions about the objective function because it is hard to imagine such a completely irregular objective function in any real physical systems. For example, under the assumpton that $f(x_i)$ and $f(x_j)$, $i \neq j$ are independent, the probability of the minimum being almost at the point of maximum is the same as at any other point.

Apparently the weakest condition which satisfies the continuity assumption, and in many cases more or less corresponds to *a priori* ideas about the pattern of behaviour of an objective function, is the independences of partial differences. The partial difference of m-th order may be regarded as an approximation of partial derivative of m-th order and can be expressed by the following recurrent formulae

$$\Delta_{\varepsilon_1}(x^1, \ldots, x^m) = f(x^1 + \varepsilon_1, \ldots, x^m) - f(x^1, \ldots, x^m)$$

$$\Delta_{\varepsilon_1 \varepsilon_2}(x^1, \ldots, x^m) = \Delta_{\varepsilon_1}(x^1, x^2 + \varepsilon_2, \ldots, x^m) - \Delta_{\varepsilon_1}(x^1, \ldots, x^m), \ldots$$

$$\ldots, \Delta_{\varepsilon_1, \ldots, \varepsilon_m}(x^1, \ldots, x^m) = \Delta_{\varepsilon_1, \ldots, \varepsilon_{m-1}}(x^1, \ldots, x^m + \varepsilon_m)$$

$$- \Delta_{\varepsilon_1, \ldots, \varepsilon_{m-1}}(x^1, \ldots, x^m),$$

$$\varepsilon_i > 0, \quad i = 1, \ldots, m. \tag{4.4.1}$$

It is well known (see Katkauskaite (1972)) that from the continuity of sample functions $f(x, \omega)$ and the independence of differences (4.4.1) it follows that the stochastic function $f(x)$ is Gaussian.

THEOREM 4.4.1 *Suppose that a Gaussian stochastic function $f(x)$ is defined on an interval $[-1, 1]^m$ with constant expectation μ, standard variance σ^2 and covariance*

$$\sigma_{jk} = \sigma^2 \prod_{i=1}^{m} \left(1 - \frac{|x_j^i - x_k^i|}{2}\right) \tag{4.4.2}$$

Then it is continuous, homogeneous and has independent partial differences.

Proof. Let us consider the Wiener field with an origin at the vertice x_l of an m-dimensional cube $[-1, 1]^m$. In such a case the covariance can be expressed as

$$\sigma_{jk}(x_l) = \sigma^2 \prod_{i=1}^{m} u_i \tag{4.4.3}$$

where

$$
u_i = \begin{cases} \min\,(x_j^i - x_l^i,\ x_k^i - x_l^i), & \text{if } x_j^i \geq x_l^i \ \text{ and } \ x_k^i \geq x_l^i \\ \min\,(x_l^i - x_j^i,\ x_l^i - x_k^i), & \text{if } x_j^i < x_l^i \ \text{ and } \ x_k^i < x_l^i \\ 0, & \text{if neither inequality is true.} \end{cases}
$$

(4.4.4)

The expectation of the Wiener field is zero and variance

$$
\sigma_x^2(x_l) = \sigma^2 \prod_{i=1}^{m} |x^i - x_l^i|.
$$

(4.4.5)

We can make the variance independent of the origin of co-ordinates x_l by summing 2^m independent Wiener fields corresponding to the different vertices of the cube $[-1, 1]^m$. Then the variance of the sum

$$
\sigma_x^2 = \frac{1}{2^m} \sum_{l=1}^{m} \sigma_x^2(x_l) = \frac{\sigma^2}{2^m} \sum_{l=1}^{2^m} \prod_{i=1}^{m} |x^i - x_l^i| = \sigma^2
$$

(4.4.6)

and the covariance of the sum

$$
\sigma_{jk} = \frac{\sigma^2}{2^m} \sum_{l=1}^{2^m} \prod_{i=1}^{m} u_i = \sigma^2 \prod_{i=1}^{m} \left(1 - \frac{|x_j^i - x_k^i|}{2}\right)
$$

(4.4.7)

This means that this stochastic function is homogeneous and can be regarded as the sum of constant μ and 2^m independent Wiener fields with origins at the vertices of the cube $[-1, 1]^m$. Since each Wiener field is a Gaussian stochastic function with independent partial differences, the sum of any finite number of independent Wiener fields is also a Gaussian stochastic function with independent partial differences.

EXAMPLE 4.4.1. If $m = 1$, $\mu = 0$, $\sigma = 1$ then vertices

$$
x_1 = -1, \ x_2 = 1
$$

and from (4.4.5)

$$\sigma_x^2(x_1) = 1 + x,$$

$$\sigma_x^2(x_2) = 1 - x.$$

From (4.4.3), (4.4.4) and (4.4.6)

$$\sigma_{jk}(x_1) = 1 + \min(x_j, x_k)$$

$$\sigma_{jk}(x_2) = 1 - \max(x_j, x_k)$$

$$\sigma_x^2 = 1/2 \, (\sigma_x^2(x_1) + \sigma_x^2(x_2)) = 1.$$

From (4.4.7)

$$\sigma_{jk} = 1/2 \, (\sigma_{jk}(x_1) + \sigma_{jk}(x_2)) = 1 - \frac{|x_j - x_k|}{2}.$$

The homogeneous Wiener field defined by (4.4.2) is a stochastic function which satisfies the conditions of continuity, homogeneity and independence of partial differences, and so it is a natural, clear and convenient stochastic model for theoretical consideration.

4.5   A case of noisy observations

In some cases the exact values of $f(x)$ cannot be defined exactly because of errors in calculations or physical experimentation. For example, errors of calculations usually arise when $f(x)$ is obtained by numerical integration of some differential equations. The errors of physical experimentation often arise in the optimal experimental design.

The problem with errors can be considered naturally by the Bayesian approach. In this case only the formulae for conditional expectation and variation should be changed.

Proper representation of different types of errors demands different stochastic models. For example, the errors of experimentation can often be adequately represented as independent Gaussian variables. However, this stochastic model is hardly acceptable for the errors of numerical integration of differential equations. They can be better represented by a stochastic function with independent partial differences such as (4.4.2).

In a general case the error functions can be considered as sample paths of different, but not necessarily independent stochastic functions. This means that the error function g can depend on two variables : on the point of observation x and its number i. Considering this case we shall denote the value of error function g_i at the point x' by $g_{i'}(x')$. The covariance between two error functions $g_{i'}$ and $g_{i''}$ at the points x' and x'' will be denoted as

$$s_{i'i''}(x', x'') = E\{g_{i'}(x'), g_{i''}(x'')\}. \tag{4.5.1}$$

It is natural to suppose that

$$E\{g_i(x)\} = 0 \tag{4.5.2}$$

and

$$s_{i'i''}(x', x'') = s(x', x'') \cdot s_{i'i''} \tag{4.5.3}$$

where $s(x', x'') = E\{g_i(x'), g_i(x'')\}$ is independent of i and $s_{i'i''} = E\{g_{i'}(x), g_{i''}(x)\}$ is independent of x.

For example, in the case of independent errors

$$s_{i'i''} = \begin{cases} 1, & i' = i'' \\ 0, & i' \neq i'' \end{cases}$$

and from (4.5.3)

$$s_{i'i''}(x', x'') = \begin{cases} s(x_i, x_i), & i' = i'' = i, \ x' = x'' = x_i \\ 0, & i' \neq i'' \end{cases}$$

In the case when the same sample path of an error function is observed (linear dependence)

$$s_{i'i''} = 1 \quad \text{and} \quad s_{i'i''}(x', x'') = s(x', x'').$$

In the case when sample paths are a Markov chain, the points of observation being the same,

$$s_{i'i''}(x', x'') = s(x', x'') (1 - a)^{|i' - i''|}, \ 0 < a < 1.$$

4.6 Estimation of parameters from dependent observations

It was shown in section 4.4 that the probability distribution P can be derived from some natural assumptions. However, the values of mean and variance remain undefined and must be estimated from observations. It is not too difficult to do this if the observations are independent. Unfortunately, it is not true if we wish to use the results of optimization including the Bayesian one for the estimation of unknown parameters because, with the exception of uniform random search, the points of optimising sequences are usually dependent.

Since we have agreed that optimality is to be understood in the average sense, it is desirable to define unbiased estimates of unknown parameters μ and σ. This is not an easy task when observations are dependent.

Let us consider the Gaussian field $f(x)$, $x \in A \subset R^m$ with mean μ and covariance $\sigma^2 \rho(s, t)$, where μ and σ are unknown and $\rho_{ij} = \rho(x_i, x_j)$ is an element of the fixed correlation matrix. An example of such a field was given in section 4.4. Suppose that $x_{i+1} = d_i(z_i)$, $i = 0,1,2, ...$, where $z_i = (x_j, f(x_j), j = 1, ... , i)$ and d_i is continuous almost everywhere.

We shall investigate the maximum likelihood estimates μ_n of unknown mean μ, where

$$\mu_n = \sum_{i,j=1}^{n} \rho_{ij}^{-1} y_i \ / \sum_{i,j=1}^{n} \rho_{ij}^{-1}, \ y_i = f(x_i) \tag{4.6.1}$$

because it is unbiased when $x_1 \quad ... , x_n$ are fixed. We can reasonably expect that this property will also occur in the case of dependent observations if they are dependent in an 'unbiased' way.

It was shown by Mockus and Senkiene (1979) that the estimate μ_n is asymptotically unbiased when

$$1/n^2 \, \text{var} \left(\sum_{i,j=1}^{n} \rho_{ij}^{-1} \right) \to 0, \ n \to \infty. \tag{4.6.2}$$

THEOREM 4.6.1. *If condition (4.6.2) holds, then the maximum likelihood estimate μ_n of mean μ defined by (4.6.1) is asymptotically unbiased*

$$E\mu_n \to \mu \ \text{when} \ n \to \infty. \tag{4.6.3}$$

Proof. Let

$$a_n = \sum_{i,j=1}^{n} \rho_{ij} y_i, \quad b_n = \sum_{i,j=1}^{n} \rho_{ij} \quad \text{and} \quad R_n^{-1} = (\rho_{ij}^{-1})$$

where R_n^{-1} can be expressed by the recurrent formula of Frobenius, see Gantmacher (1967)

$$R_n^{-1} = \begin{pmatrix} R_{n-1}^{-1} + R_{n-1}^{-1} \begin{pmatrix} \rho_{n\,1} \\ \vdots \\ \rho_{n\,n-1} \end{pmatrix} B^{-1}(\rho_{n\,1}, \ldots, \rho_{n\,n-1}) R_{n-1}^{-1} & -R_{n-1}^{-1} \begin{pmatrix} \rho_{n\,1} \\ \vdots \\ \rho_{n\,n-1} \end{pmatrix} B^{-1} \\[20pt] -B^{-1}(\rho_{n\,1}, \ldots, \rho_{n\,n-1}) R_{n-1}^{-1} & B^{-1} \end{pmatrix}$$

Here

$$B = \rho_{n\,n} - (\rho_{n\,1}, \ldots, \rho_{n\,n-1}) R_{n-1}^{-1} \begin{pmatrix} \rho_{n\,1} \\ \vdots \\ \rho_{n\,n-1} \end{pmatrix}$$

$$a_n = a_{n-1} + 1/H \left((\rho_{n\,1}, \ldots, \rho_{n\,n-1}) R_{n-1}^{-1} \begin{pmatrix} 1 \\ \vdots \\ 1 \end{pmatrix} - 1 \right)$$

$$\times \left((\rho_{n\,1}, \ldots, \rho_{n\,n-1}) R_{n-1}^{-1} \begin{pmatrix} y_1 \\ \vdots \\ y_{n-1} \end{pmatrix} - y_n \right)$$

$$b_n = b_{n-1} + 1/H \left((1, \ldots, 1) R_{n-1}^{-1} \begin{pmatrix} \rho_{n\,1} \\ \vdots \\ \rho_{n\,n-1} \end{pmatrix} - 1 \right)^2$$

$$H = \rho_{n\,n} - (\rho_{n\,1}, \,\cdots\,, \rho_{n\,n-1})\, R_{n-1}^{-1} \begin{pmatrix} \rho_{n\,1} \\ \vdots \\ \rho_{n\,n-1} \end{pmatrix}$$

Since d_n, $n = 0, 1, \ldots, N$ are assumed to be continuous, then from Theorem 2.4.1 it follows that the conditional expectation and variance of $f(x)$ in the case of dependent observations can be expressed using the same formulae (4.3.3) and (4.3.4) as for the fixed points $x_1, x_2, \ldots, x_n$. In such a case the relation between the conditional expectation of a_n and that of b_n

$$\frac{Ea_n}{Eb_n} = \mu \; \frac{E\left\{1/H\left((1, \ldots, 1)\, R_{n-1}^{-1} \begin{pmatrix} \rho_{n\,1} \\ \vdots \\ \rho_{n\,n-1} \end{pmatrix} - 1\right)^2\right\} + \ldots + E\left\{1/\rho_{11}\right\}}{E\left\{1/H\right\}\left((1, \ldots, 1)\, R_{n-1}^{-1} \begin{pmatrix} \rho_{n\,1} \\ \vdots \\ \rho_{n\,n-1} \end{pmatrix} - 1\right)^2 + \ldots + E\left\{1/\rho_{11}\right\}} = \mu$$

It follows from condition (4.6.2) and the Markov theorem, see Gnedenko (1965), that for any $\varepsilon > 0$

$$\lim_{n \to \infty} P\left\{\left|1/n \sum_{i,j=1}^{n} \rho_{ij}^{-1} - 1/n \sum_{i,j=1}^{n} E\rho_{ij}^{-1}\right| < \varepsilon\right\} = 1.$$

Since

$$\left|\frac{\sum\limits_{i,j=1}^{n} \rho_{ij}^{-1}\, y_i}{\sum\limits_{i,j=1}^{n} \rho_{ij}^{-1}}\right| \leq \frac{\sum\limits_{i,j=1}^{n} \rho_{ij}^{-1}\, |y_i|}{\sum\limits_{i,j=1}^{n} \rho_{ij}^{-1}} \leq \left(\sum\limits_{i,j=1}^{n} y_i^2\right)^{1/2}$$

then it follows from the Lebesgue convergence theorem that

$$E\mu_n = E \; \frac{1/n \; \sum\limits_{i,j=1}^{n} \rho_{ij}^{-1} \; y_i}{1/n \; \sum\limits_{i,j=1}^{n} \rho_{ij}^{-1}} \quad \rightarrow \quad \frac{E \sum\limits_{i,j=1}^{n} \rho_{ij}^{-1} \; y_i}{E \sum\limits_{i,j=1}^{n} \rho_{ij}^{-1}} = \mu$$

when $n \rightarrow \infty$.

To test condition (4.6.2), we shall define some more simple conditions from which condition (4.6.2) follows.

Suppose there exists a converse correlation matrix and it is differentiable.

$$p_{ij} = \rho_{ij}^{-1}, \tag{4.6.4}$$

$$\rho_{ij}^l = \partial \rho_{ij}/\partial x^l, \, l = 1, \dots , m. \tag{4.6.5}$$

Suppose that p_{ij}^l are bounded, where

$$p_{ij}^l = \partial p_{ij}/\partial x^l, \, l = 1, \dots , m. \tag{4.6.6}$$

and that there exists a fixed sequence $\tau_n \in A, \, n = 1, \dots N$ where to each n there corresponds index i_n such that

$$\|\tau_n - x_{i_n}\| < \varepsilon_N, \quad \varepsilon_N > 0, \quad \varepsilon_N \rightarrow 0 \text{ when } N \rightarrow \infty. \tag{4.6.7}$$

The purpose of the last condition is to reduce the difference between the fixed and the random points to the ε-level, where $\varepsilon \rightarrow 0$ when $N \rightarrow \infty$.

THEOREM 4.6.2. *Condition (4.6.2) follows from conditions (4.6.4) to (4.6.7).*

Proof. For any k, s the difference

$$p_{i_k j_s} - p_{ks} = \sum\limits_{l=1}^{m} \left(\left(\Delta_k^l + \Delta_s^l \right) h_{ks}^l + o\left(\Delta_k^l \right) + o\left(\Delta_s^l \right) \right) \tag{4.6.8}$$

where

$$p_{ks} = \rho^{-1}(\tau_k, \tau_s), \quad p_{i_k j_s} = \rho^{-1}(x_{i_k}, x_{j_s}),$$

$$\Delta_k^l = |\tau_k^l - x_{i_k}^l|, \quad h_{ks}^l = (\partial p_{i_k j_s})/\partial x^l.$$

Since any $k = 1, \dots, N$ corresponds to one and only one i_k, then

$$\sum_{k,s=1}^{N} p_{i_k j_s} = \sum_{i,j=1}^{n} p_{ij}, \quad \text{where } p_{ij} = \rho^{-1}(x_i, x_j)$$

From here and (4.6.8) it follows that

$$\sum_{i,j=1}^{N} p(x_i, x_j) - \sum_{k,s=1}^{N} p(\tau_k, \tau_s) = 2 \sum_{k=1}^{N} \sum_{l=1}^{m} \Delta_k^l \left(\sum_{s=1}^{N} h_{ks}^l + o(\Delta_s^l) \right).$$

Omitting the remainder

$$\sum_{i,j=1}^{N} p(x_i, x_j) - \sum_{k,s=1}^{N} p(\tau_k, \tau_s) = 2 \sum_{k=1}^{N} \sum_{l=1}^{m} \Delta_k^l h_k^l$$

where $h_k^l = \sum\limits_{s=1}^{N} h_{ks}^l$.

Hence

$$\sum_{i,j=1}^{N} p(x_i, x_j) = \sum_{k,s=1}^{N} p(\tau_k, \tau_s) + 2 \sum_{k=1}^{N} \sum_{l=1}^{m} \Delta_k^l h_k^l \tag{4.6.9}$$

Because $\sum\limits_{k,s=1}^{N} p(\tau_k, \tau_s)$ is fixed, the variance of sum (4.6.9) can be expressed as

$$\text{var} \left(\sum_{i,j=1}^{N} p_{ij} \right) = 4 \, \text{var} \sum_{k=1}^{N} \sum_{l=1}^{m} \xi_k^l \leq 4N^2 m^2 C \tag{4.6.10}$$

where

$$\xi_k^l = \Delta_k^l h_k^l, \quad C = \max_{\substack{k,s=1,\dots,N \\ l=1,\dots,m}} |E(\xi_k^l - E\xi_k^l)(\xi_s^l - E\xi_s^l)| \qquad (4.6.11)$$

Suppose that from

$$\Delta = \max_{\substack{k=1,\dots,N \\ l=1,\dots,m}} \Delta_k^l \to 0$$

it follows

$$C \to 0. \qquad (4.6.12)$$

Then

$$1/N^2 \operatorname{var}\left(\sum_{i,j=1}^{N} p_{ij}\right) \to 0 \qquad (4.6.13)$$

when $N \to \infty$.

(4.6.12) is true because for any random $y_1, y_2 \in [-\Delta, \Delta]$ the following inequality holds

$$E\{(y_1 - Ey_1)(y_2 - Ey_2)\} \le \Delta^2$$

and from (4.6.7) it follows that $\Delta \to 0$.

Condition (4.6.7) follows from Lemma 4.2.6 and the definition (4.2.19).

COROLLARY 4.6.3. *If there exists an inverse correlation matrix R_n^{-1} with bounded derivatives and the conditions of Lemma 4.2.6 hold, then the estimate (4.6.1) is asymptotically unbiased.*

So far we have been considering only the estimate of the mean. Now let us consider the maximum likelihood estimate σ_n^2 of variance σ^2

$$\sigma_n^2 = \frac{1}{N-1} \sum_{i,j=1}^{N} \rho_i^{-1} v_i v_j. \qquad (4.6.14)$$

Here

$$v_i = f(x_i) - \mu_n, \quad v_j = f(x_j) - \mu_n.$$

If the observations are fixed then the estimate (4.6.14) is unbiased. If condition (4.6.2) holds, σ_n^2 can be expected to be nearly unbiased for the same reasons as in the case of the mean, also in the case of dependent observations.

CHAPTER 5

BAYESIAN METHODS FOR GLOBAL OPTIMIZATION IN THE GAUSSIAN CASE

5.1 The one-step approximation

The formula for the one-step Bayesian approach assuming the Gaussian distribution is from (2.5.1) and (2.5.5)

$$x_{n+1} \in \arg\min_{x \in A} (1/\sigma) \int_{-\infty}^{\infty} \min(y, c) \exp\left((-1/2)((y-\mu)/\sigma)^2\right) dy. \qquad (5.1.1)$$

Here μ is the conditional expectation and σ is the conditional standard deviation of $f(x)$ when the observed values are

$$y_1 = f(x_1), \dots, y_n = f(x_n)$$

and

$$c = \min_{x \in A} \mu - \varepsilon, \quad \varepsilon > 0. \qquad (5.1.2)$$

Here ε takes into account the influence of subsequent observations. When ε is large, the method becomes a nearly uniform search. When ε approaches zero the method is strictly one-step.

The relation (5.1.1) can be expressed as

$$x_{n+1} \in \arg\max_{x \in A} (1/\sigma) \int_{-\infty}^{c} (c-y) \exp\left((-1/2)((y-\mu)/\sigma)^2\right) dy. \qquad (5.1.3)$$

Let

$$u = (y-\mu)/\sigma \qquad (5.1.4)$$

and

$$a = (c-\mu)/\sigma. \qquad (5.1.5)$$

Parameter $a < 0$, because, as follows from (5.1.2), $c < \mu$. From (5.1.4), (5.1.5) and (5.1.3) we can write

$$x_{n+1} \in \arg \max_{x \in A} \phi(a) \tag{5.1.6}$$

where

$$\phi(a) = \sigma \int_{-\infty}^{a} (a - u) \exp((-1/2)u^2)\, du. \tag{5.1.7}$$

The derivative of ϕ with respect to a

$$\frac{d\phi}{da} = \frac{\partial \phi}{\partial a} + \frac{\partial \phi}{\partial \sigma} \frac{d\sigma}{da} \tag{5.1.8}$$

where

$$\partial\phi/\partial a = \partial/\partial a \left(\sigma a \int_{-\infty}^{a} \exp((-1/2)u^2)\, du - \sigma \int_{-\infty}^{a} u \exp((-1/2)u^2)\, du \right.$$

$$= \sigma \int_{-\infty}^{a} u \exp((-1/2)u^2)\, du. \geq 0 \text{ because } \sigma \geq 0 \tag{5.1.9}$$

and

$$\partial\phi/\partial\sigma = \int_{-\infty}^{a} (a - u) \exp((-1/2)u^2)\, du > 0 \tag{5.1.10}$$

because $a - u > 0$ for all $u \in (-\infty, a)$

$$\frac{d\sigma}{da} = \frac{d}{da}\left(\frac{c - \mu}{a}\right) = \frac{\sigma^2}{\mu - c} > 0. \tag{5.1.11}$$

It follows from (5.1.8), (5.1.9), (5.1.10) and (5.1.11) that

$$d\phi/da > 0 \tag{5.1.12}$$

because

$$d\phi/da \geq 0, \quad \partial\phi/\partial\sigma > 0 \text{ and } d\sigma/da > 0. \tag{5.1.13}$$

The inequality (5.1.12) means that ϕ is an increasing function of a, so from (5.1.6)

$$x_{n+1} \in \arg\max_{x \in A} a = \arg\max_{x \in A} (-1/a) = \arg\max_{x \in A} \phi(x) \qquad (5.1.14)$$

Here

$$\phi(x) = \sigma/(\mu - c). \qquad (5.1.15)$$

The maximum of ϕ can be either on the boundary of A or at the point where the differential of ϕ is zero

$$\frac{d\phi}{dx} = \frac{\partial\phi}{\partial\sigma}\frac{d\sigma}{dx} + \frac{\partial\phi}{\partial\mu}\frac{d\mu}{dx} = \frac{\partial\phi}{\partial\sigma}\sigma' + \frac{\partial\phi}{\partial\mu}\mu'$$

$$= \sigma'/(\mu - c) - \mu'\sigma/(\mu - c)^2 = 0. \qquad (5.1.16)$$

It follows from (5.1.16) that

$$\sigma'/\sigma = \mu'/(\mu - c). \qquad (5.1.17)$$

Solution of (5.1.14) or (5.1.17) is rather time consuming in a multi-dimensional case. No more that 100 to 200 observations can be handled when expressions for μ and σ correspond to the usual multi-dimensional Gaussian distribution.

It appears that no further substantial simplifications can be made if we wish to satisfy the Kolmogorov consistency conditions. These consistency conditions ensure that a sample path of the same stochastic function is considered during the process of optimization. Some additional requirements given by Katkauskaite (1975) allow us to consider only continuous sample paths.

It is easy to notice that the consistency conditions can make the stochastic model less adaptive. For example, if we update the parameters μ_0 and σ_0 of a stochastic function on the basis of the results of observations, we violate the consistency conditions. However, this usually improves the results of the optimization because it adapts the stochastic model to the observed data, so it is reasonable to omit those conditions if by doing so we get some computational advantage and if convergence remains.

5.2 Adaptive models

There are many ways to simplify the stochastic model (5.1.1) if the usual consistency conditions are dropped. Let us consider the situation when the conditions of consistency and continuity of sample paths are omitted but the conditions of convergence of method and continuity of the function $\phi(x)$, the maximum which defines the next point of observation in accordance with (5.1.14), remain.

The adaptive model will be defined in the following way.

$$f(x) = f_i(x); \quad x \in A_i; \quad \bigcup_{i=1}^{n} A_i = A; \quad A_i \cap A_j = \varnothing, \quad i \neq j; \quad i, j = 1, \dots, n$$

$$(5.2.1)$$

where each A_i contains one observation x_i. Suppose that $f_i(x)$ is a Gaussian stochastic function, 'conditional' expectation μ_x^i is equal to the observed value y_i and 'conditional' variance $\sigma_x^{i^2}$ is an increasing function Δ of the distance $d_i = \|x - x_i\|$ from the point of observation x_i, namely

$$\mu_x^i = y_i \quad \text{and} \quad \sigma_x^{i^2} = \Delta(d_i), \quad x \in A_i. \tag{5.2.2}$$

It follows from (5.1.15) and (5.2.2) that

$$\phi_i(x) = \sigma_x^i / (\mu_x^i - c) \tag{5.2.3}$$

and

$$\phi(x) = \phi_i(x), \quad x \in A_i. \tag{5.2.4}$$

Since $\phi_i(x)$, $i = 1, \dots, n$ are fixed by (5.2.3), the postulated continuity of $\phi(x)$ can be provided only by the proper choice of A_i, namely when

$$A_i = \{x : \phi_i(x) \leq \phi_j(x), \ j = 1, \dots, n\}. \tag{5.2.5}$$

From (5.2.4) and (5.2.5)

$$\phi(x) = \min_{1 \leq i \leq n} \phi_i(x).$$

From this and from (5.1.14)

$$x_{n+1} \in \arg \max_{x \in A} \ \min_{1 \le i \le n} \ \phi_i(x) \tag{5.2.6}$$

or taking (5.2.3) into consideration

$$x_{n+1} \in \arg \max_{x \in A} \ \min_{1 \le i \le n} \ \sigma_x^i /(\mu_x^i - c). \tag{5.2.7}$$

Most of the calculations so far have been done using a Gaussian stochastic function with conditional expectation (5.2.2) and the conditional variance

$$\sigma_x^{i^2} = \sigma_0^2 \|x - x_i\|^4 \tag{5.2.8}$$

because the model is simplest in this case. Method (5.2.7) satisfies the conditions of Theorem 4.2.9, sufficient for the convergence of the sequence (5.2.7) to the global minimum of any continuous function. This means that the use of an adaptive model (5.2.2) and (5.2.7) developed without the conditions of consistency of distribution functions and of continuity of sample paths shows the same asymptotic results as the standard one-step stochastic model (5.1.1), which is consistent and continuous but more complicated. The results of calculations have not shown any substantial difference between the standard and adaptive models except reduction of computational effort.

So far the adaptive model (5.2.7) has been regarded as a simplification of the standard one-step Bayesian model (5.1.1) which was supposed to be an approximation of the classical probabilistic Bayesian model (2.1.11) represented in section 2.2 as the system of recurrent equations (2.2.1) of dynamic programming.

It seems, however, much more interesting and useful to consider the adaptive model (5.2.7) as a different kind of stochastic model, which may correspond even better to the basic ideas of Bayesian decision theory in the problem of global optimization. The classical probabilistic Bayesian model (2.1.11), (2.2.1) is based on assumptions directly borrowed from the classical theory of stochastic functions, such as Kolmogorov's consistency conditions. However, the conditions are not so important in some Bayesian decision problems, including global optimization, when it is not considered necessary to see the objective function as a sample path of a fixed stochastic function.

So it would be better to regard the adaptive Bayesian model (5.2.7) as some non-classical stochastic model where the usual consistency conditions are replaced by the condition of the continuity of risk function (5.1.1) and the expressions of

'conditional' expectation and 'conditional' variance are made as simple as possible
ensuring the convergence to the minimum of any continuous function $f(x)$.

ASSUMPTION 5.2.1. The simplicity of functions corresponds to the following
ordering

    1) constant function
    2) step function
    3) linear function
    4) piecewise-linear function
    5) quadratic function
    6) piecewise-quadratic function.

DEFINITION 5.2.1. A non-negative function

$$p_x(y) = p_x(y/x_i, y_i, \; i = 1, \dots, n), \; x \in A, \; y \in R, \; x_i \in A, \; y_i \in R,$$

will be called a non-consistent conditional density function, if

$$\int_{-\infty}^{\infty} p_x(y) \, dy = 1.$$

The term 'non-consistent' is necessary to make the distinction from the usual
definition of conditional density which satisfies the consistency conditions. In this
section and later, the term 'non-consistent' is omitted if it does not lead to confusion
with the classical definition of conditional probabilities.

THEOREM 5.2.1. *Suppose that*

    1) *The conditional density function* $p_x(y)$ *is Gaussian with conditional mean* μ_x
and conditional variance σ_x^2.

    2) *The risk function* (5.1.1) *is continuous.*

    3) *The sequence* (5.1.1) *satisfies the convergence conditions* (4.2.65).

    4) *The conditional mean* μ_x *and the conditional standard deviation* σ_x *are
functions of x, as simple as possible in the sense of Assumption 5.3.1.*

Then the one-step Bayesian method is defined by expressions (5.2.7) and (5.2.8).

Proof. The simplest function which satisfies conditions $(4.2.26^b)$ is the step function $\mu_x = f(x_i)$, $x \in A_i$ where A_i, $i = 1, ... , n$ is the partition of A such that $x_i \in A_i$.

The simplest function which satisfies conditions (4.2.30) and (4.2.31) is the step function : $\sigma_x = 0$ if $x = x_i$ and $\sigma_x = h > 0$ if $x \neq x_i$, $i = 1, ... , n$ but this function contradicts the condition of continuity of risk function. To satisfy this condition the conditional standard deviation σ_x should be an increasing function of the distance $\|x - x_i\|$. The simplest one in the sense of assumption 5.2.1 being $\sigma_x = \|x - x_i\|^2$. In this case the risk function (4.2.35) and the corresponding function $\phi(x)$, see (5.1.15), will be continuous if the partition A_i, $i = 1, ... , n$ corresponds to the condition (5.2.5). The remaining conditions of Lemma 4.2.6 are also satsisfied. So in this case the convergence condition holds, if there is no noise.

It follows from Theorem 4.2.10 that in the case of noisy observations the convergence of the Bayesian method does not necessarily follow from the condition of asymptotic density (4.2.65), as it does in the absence of noise. In the presence of noise the convergence of the method is provided if the conditional expectation converges to the function to be minimized, see condition (4.2.36). It is obviously not true in the case (5.2.2), (5.2.7). So in the noisy case the standard Bayesian method (5.1.1) should be used instead of the adaptive one (5.2.7) if the convergence of the method is necessary. If not, then the simpler method (5.2.7) can be used which does not necessarily converge but provides some improvement in the average sense of objective function with regard to the initial point. If the convergence is needed but the standard Bayesian method is considered to be too complicated, the following algorithm can be applied, see Mockus et al (1987).

1) The observations are restricted by the condition

$$
y_i = \begin{cases} h_{x_i}, & \text{if } h_{x_i} \geq y_0 \\[2mm] y_0, & \text{if } h_{x_i} < y_0 \end{cases}
$$

where y_0 is some acceptable level and $h_{x_i} = f(x_i) + \xi_i$

2) The point of the next observation is defined by the adaptive Bayesian method (5.2.7).

3) The final decision x_{N+1} is defined using some statistical model which corresponds to the condition (4.2.36).

5.3 Extrapolation models

The conditional expectation and conditional variance can also be considered in the framework of the theory of extrapolation under uncertainty, see Zilinskas (1982). There are five assumptions concerning conditional expectation μ_x^k :

ASSUMPTION 5.3.1.

$$\mu_x^k((x_i, cy_i), \; i = 1, \ldots, k) \; = \; c\mu_x^{\,k}((x_i, y_i), \; i = 1, \ldots, k)$$

for any real c.

ASSUMPTION 5.3.2.

$$\mu_x^k((x_i, y_i + c), \; i = 1, \ldots, k) \; = \; \mu_x^{\,k}((x_i, y_i), \; i = 1, \ldots, k) + c$$

for any real c.

ASSUMPTION 5.3.3.

$$\mu_x^k((x_i, y_i), \; i = 1, \ldots, k) \; = \; \mu_x^{\,k}((x_{j(i)}, y_{j(i)}), \; i = 1, \ldots, k)$$

for any permutation of indices $j(i)$.

ASSUMPTION 5.3.4.

$$\mu_{x_j}^k((x_i, y_i), \; i = 1, \ldots, k) \; = \; y_j, \; j = 1, \ldots, k.$$

ASSUMPTION 5.3.5. There exists a mapping $v_k(\cdot) : A \times (A \times R)^{k-1} \to R$ such that

$$\mu_x^k((x_i, y_i), \ i = 1, \ldots, k) = \mu_x^k((x_i, v_i), \ i = 1, \ldots, k)$$

where

$$v_i = v, \ i \neq k \quad \text{and} \quad v_k = y_k.$$

Here

$$v = v_k(x, (x_i, y_i), \ i = 1, \ldots, k-1).$$

Assumptions 5.3.1 to 5.3.4 look fairly natural. Assumption 5.3.5 restricts us to the situations when the results of all $k-1$ observations can be aggregated into one single number v.

THEOREM 5.3.1. *The unique function satisfying assumptions 5.3.1 to 5.3.5 is the weighted sum* :

$$\mu_x^k((x_i, y_i), \ i = 1, \ldots, k) = \sum_{i=1}^{k} y_i s_i^k (x, x_j, \ j = 1, \ldots, k) \qquad (5.3.6)$$

where weights s_i^k possess the following properties

$$\sum_{i=1}^{k} s_i^k (x, x_j, \ j = 1, \ldots, k) = 1. \qquad (5.3.7)$$

$$s_i^k(x, x_j, \ j = 1, \ldots, k)$$

$$= s_p^k(x, x_1, \ldots, x_{j-1}, x_l, x_{j+1}, \ldots, x_{l-1}, x_j, x_{l+1}, \ldots, x_k). \qquad (5.3.8)$$

Here

$$p = \begin{cases} i, & \text{if } i \neq j, \ i \neq l, \\ l, & \text{if } i = j, \\ j, & \text{if } i = l \end{cases}$$

and

$$s_i^k(x_l, x_j, \ j = 1, \ldots, k) = \begin{cases} 1, & i = l \\ 0, & i \neq l \end{cases} \qquad (5.3.9)$$

Proof. Suppose that $x, x_i, i = 1, \ldots, k$ are fixed. Denote

$$\mu_x^k((x_i, v), \ i = 1, \ldots , k-1, (x_k, v)) \ = \ \phi_k(u, v).$$

It follows from assumptions 5.3.1, 5.3.2 that

$$\phi_k(cu, cv) \ = \ c\,\phi_k(u, v)$$

and

$$\phi_k(u + c, v + c) \ = \ \phi_k(u, v) + c$$

in accordance with Aczel (1966).

$$\phi_k(u, v) \ = \ \frac{a_k u + b_k v}{a_k + b_k}$$

Hence from assumption 5.3.5

$$\mu_x^k((x_j, y_j), \ j = 1, \ldots , k) \ = \ \frac{a_k v_k(x, (x_j, y_j), \ j = 1, \ldots , k-1) + b_k y_k}{a_k + b_k}$$

$$(5.3.10)$$

Because of the independence of permutations (assumption 5.3.3)

$$\mu_x^k((x_j, y_j), \ j = 1, \ldots , k) \ = \ \frac{a_l v_l(x, (x_j, y_j), \ j = 1, \ldots , k, \ j \neq l) + b_l y_{l l}}{a_l + b_l}$$

$$l = 1, \ldots , k-1. \qquad\qquad\qquad\qquad\qquad\qquad\qquad (5.3.11)$$

It follows from (5.3.10) and (5.3.11) that μ_x^k is differentiable with respect to y_l for any fixed $l, l = 1, \ldots , k$ and

$$\partial/\partial y_l \, \mu_x^k((x_j, y_j), \ j = 1, \ldots , k) \ = \ \frac{b_l}{a_l + b_l} \ , \ l = 1, \ldots , k. \qquad (5.3.12)$$

Therefore

$$\mu_x^k((x_j, y_j), \ j = 1, \ldots , k) \ = \ \sum_{l=1}^{k} \frac{b_l}{a_l + b_l} \ y_l \,. \qquad\qquad (5.3.13)$$

Denote

$$\frac{b_j}{a_j + b_j} = s_j^k (x, x_i, \ i = 1, \dots, k) \tag{5.3.14}$$

and we shall have expression (5.3.6).

It follows from assumptions 5.3.1 and 5.3.2 that

$$\mu_x^k((x_j, y), \ j = 1, \dots, k) = y. \tag{5.3.15}$$

Therefore

$$\sum_{j=1}^{k} s_j^k (x, x_i, \ i = 1, \dots, k) = 1.$$

Properties (5.3.8) and (5.3.9) of the weights s_i^k follow directly from assumptions 5.3.3 and 5.3.4, respectively.

Now consider the following seven assumptions concerning conditional variance $\sigma_{xx}^k = s^k(x, (x_i, y_i), \ i = 1, \dots, k)$.

ASSUMPTION 5.3.16. There exists a mapping $\sigma(x, z)$, $A^2 \to R$ such that

$$\sigma(x, z) = \sigma(z, x), \ \sigma(x, x) > \sigma(x, z), \ x \neq z$$

and

$$s^k(x, (x_i, y_i), \ i = 1, \dots, k) = \gamma_k S_k(\sigma, (\sigma_i, s_i^{\ k}), \ i = 1, \dots, k)$$

where

$$\sigma = \sigma(x, x), \ \sigma_i = \sigma(x, x_i), \ \gamma_k = \gamma_k(y_1, \dots, y_k),$$

$$s_i^k = s_i^k(x, x_j, \ j = 1, \dots, k).$$

Here γ_k depends only on the results of observations $y_1, \dots, y_k$ and S_k depends only on the weights s_i^k and the functions σ, σ_i which do not depend on $y_1, \dots, y_k$.

ASSUMPTION 5.3.17.

$$S_k(\sigma + \sigma', (\sigma_i, s_i^{\ k}), i = 1, \dots, k) = S_k(\sigma, (\sigma_i, s_i^{\ k}), i = 1, \dots, k) + \sigma'.$$

ASSUMPTION 5.3.18.

$$S_k(\sigma + \sigma', (\sigma_i + \sigma', s_i^{\ k}), i = 1, \dots, k) = S_k(\sigma, (\sigma_i, s_i^{\ k}), i = 1, \dots, k).$$

ASSUMPTION 5.3.19.

$$S_k(c\sigma, (c\sigma_i, s_i^{\ k}), i = 1, \dots, k) = cS_k(\sigma, (\sigma_i, s_i^{\ k}), i = 1, \dots, k).$$

ASSUMPTION 5.3.20.

$$S_k(\sigma, (\sigma_{j(i)}, s^k_{\ j(i)}), i = 1, \dots, k) = S_k(\sigma, (\sigma_i, s_i^{\ k}), i = 1, \dots, k)$$

for any permutation of indices

$$\{j(i), i = 1, \dots, k\} = \{1, 2, \dots, k\}.$$

ASSUMPTION 5.3.21. There exists a function $u_k(\cdot) : R \times (R^2)^{k-1} \to R$ such that

$$S_k(\sigma, (\sigma'_i, s_i^k), i = 1, \dots, k) = S_k(\sigma, (\sigma_i, s_i^k), i = 1, \dots, k)$$

where

$$\sigma'_i = u, \ i \neq k, \ \sigma'_k = \sigma_k.$$

Here

$$u = u_k(\sigma, (\sigma_i, s_i^{\ k}), i = 1, \dots, k-1).$$

ASSUMPTION 5.3.22.

$$S_k(\sigma, (\sigma''_i, s''^k_i), i = 1, \dots, k) = S_k(\sigma, (\sigma_i, s_i^{\ k}), i = 1, \dots, k)$$

where

$$s''^k_i = s_i^k, \ i \neq j, \ i \neq k, \ \sigma''_i = \sigma_i, \ i \neq j, \ \sigma_k = 0$$

and

$$s''^k_j \sigma''_j = s_j^k \sigma_j.$$

Assumption 5.3.16 means that the conditional variance σ_{xx}^k of $f(x)$ when $f(x_i)$, $i = 1, \dots, k$ are observed is a product of γ_k and S_k, where γ_k depends only on $y_i = f(x_i)$, $i = 1, \dots, k$ and S_k depends only on $\sigma, \sigma_i, s_i^k, i = 1, \dots, k$.

The difference

$$\delta_i = \sigma - \sigma_i = \sigma(x, x) - \sigma(x, x_i)$$

can be regarded as an uncertainty of $f(x)$ when $f(x_i)$ is observed. So σ_i may be considered as 'information' on $f(x)$ obtained observing $f(x_i)$ and σ may be regarded as 'information' on $f(x)$ obtained observing $f(x)$.

According to the assumptions 5.3.17 and 5.3.18, conditional variance σ_{xx}^k of $f(x)$ will increase by σ' if information σ increases by σ'. Correspondingly σ_{xx}^k will not change if information σ and σ_i, $i = 1, \dots, k$ are increased by the same level σ'. The assumptions look fairly natural because in the case of assumption 5.3.17 the uncertainty δ_i of $f(x)$ will be increased by σ' and in the case of assumption 5.3.18 the uncertainty δ_i will remain the same.

Assumption 5.3.21 is similar to assumption 5.3.5 and gives a possibility of data aggregation. Assumption 5.3.22 states that the conditional variance will remain the same in spite of changing weights s_j^k and information σ_j if the product $s_j^k \sigma_j$ does not change. This means that the weight s_i^k and information σ_i play an equally important role in defining the conditional variance.

THEOREM 5.3.2. *The unique function satisfying assumptions* 5.3.11 *to* 5.3.22 *is a weighted sum of uncertainty* δ_i

$$\sigma_{xx}^k = s^k(x, (x_i, y_i), i = 1, \dots, k) = \gamma_k \sum_{i=1}^k \delta_i s_i^k \qquad (5.3.23)$$

where

$$\delta_i = \sigma(x, x) - \sigma(x, x_i) \qquad (5.3.24)$$

and

$$s_i^k = s_i^k(x, x_j), \ j = 1, \dots, k) \qquad (5.3.25)$$

Proof. It follows from 5.3.17 that

$$S_k (\sigma, (\sigma_i, s_i^k), i = 1, \dots, k) = \sigma - T_k ((\sigma_i, s_i^k), i = 1, \dots, k).$$

Comparing assumptions 5.3.18 to 5.3.21 with the corresponding assumptions 5.3.1 to 5.3.3. and 5.3.5 it is easy to see that the properties of the function

$$T_k \left((\sigma_i, s_i^k), \, i = 1, \, \dots \, , k \right) \; = \; S_k \left(0, \, (\sigma_i, s_i^k), \, i = 1, \, \dots \, , k \right)$$

as a function of σ_i are the same as those for μ_x^k as a function of y_i if T_k and σ_i are substituted for μ_x^k and y_i respectively.   So the conditional variance can be represented as a weighted sum (5.3.23) of uncertainties δ_i expressed in the same way as the conditional expectation was expressed as a weighted sum (5.3.6).

The assumptions 5.3.1 to 5.3.5 and 5.3.16 to 5.3.22 are not restrictive enough since almost any reasonable expressions of conditional expectation and conditional variance can be represented as (5.3.6) and (5.3. 24) respectively.
In order to define the model more precisely, we must introduce some additional assumptions, for example the following consistency conditions.

ASSUMPTION 5.3.26. If

$$v_i \; = \; y_i, \; i = 1, \, \dots \, , k - 1$$

and

$$v_k \; = \; \mu_{x_k}^{k-1} \left((x_i, y_i), \, i = 1, \, \dots \, , k - 1 \right)$$

then

$$\mu_x^k \left((x_i, v_i), \, i = 1, \, \dots \, , k \right) \; = \; \mu_x^{k-1} \left((x_i, y_i), \, i = 1, \, \dots \, , k - 1 \right).$$

ASSUMPTION 5.3.27. Function $\sigma(x, z)$ defined in assumption 5.3.16 is positive definite and

$$\mu_x^2 \left((x_i, \sigma(z, x_i)), \, i = 1, 2 \right) \; = \; \mu_z^2 \left((x_i, \sigma(x, x_i)), \, i = 1, 2 \right)$$

for any $x_1, x_2 \in A$.

Assumption 5.3.26 expresses the compatability of μ^{k-1} and μ^k. This means that the conditional expectation will not change if the result of observation v_k is equal to the value of conditional expectation at the observed point x_k.

Assumption 5.3.27 implies a sort of symmetry which means that the conditional expectation at the point $x \in A$ with regard to the observed values

$f(x_i) = \sigma(z, x_i)$, $i = 1, 2$ is equal to the conditional expectation at the point $z \in A$ with regard to the observed values $f(x_i) = \sigma(x, x_i)$, $i = 1, 2$.

THEOREM 5.3.3. *If assumptions 5.3.26 and 5.3.27 hold, then the expressions (5.3.6) and (5.3.23) correspond to those of conditional mean and conditional variance of the Gaussian stochastic function $f(x)$.*

$$\mu_x^k \left((x_i, y_i), i = 1, \ldots, k\right) = (y_1, \ldots, y_k) \ \sum\nolimits_k^{-1} (\sigma_1, \ldots, \sigma_k)^T$$

and

$$s^k (x, (x_i, y_i), i = 1, \ldots, k) = \sigma - (\sigma_1, \ldots, \sigma_k) \ \sum\nolimits_k^{-1} (\sigma_1, \ldots, \sigma_k)^T$$

where Σ_k is a matrix with the elements $\sigma(x_i, x_j)$, $i, j = 1, \ldots, k$.

The proof is given by Zilinskas (1979).

Theorem 5.3.3 shows that the price we must pay for the consistency conditions (5.3.26) and (5.3.27) is the computation of the inverse Σ_k^{-1} of matrix Σ_k with elements $\sigma(x_i, x_j)$, $i, j = 1, \ldots, k$. It can be too big if the number of observations k is large. Besides, the consistency conditions are not necessary when adaptive models are considered, the distribution function being allowed to change in accordance with new data.

The idea of Zilinskas for resolving the contradiction is to choose weights from heuristic considerations, some of which were suggested by Shepard (1965)

$$s_i^k = s_i^k (x, x_j, j = 1, \ldots, k) = d(x, x_i) / \sum_{j=1}^{k} d(x, x_j) \qquad (5.3.28)$$

where

$$d(x, x_i) = \exp (- c \|x - x_i\|^2) / \|x - x_i\|, \ c > 0.$$

It is recommended by Zilinskas (1972) to use $c = 3.3$, if the scales in R^n are defined by normalization of the components of x by the standard deviations of the corresponding components of vectors x_i, $i = 1, \ldots, k$, (see Zilinskas (1978)).

Another idea, that by Mockus, was described in section 5.2 of this book, where the adaptive Bayesian models were considered. There it was supposed that

$$s_i = \begin{cases} 1, & x \in A_i \\ 0, & x \bar{\in} A_i \end{cases}$$

where A_i was uniquely defined by a very natural condition of continuity of the risk function.

5.4 Maximum likelihood models

The method of maximum likelihood is well known and widely used in mathematical statistics. So it is quite natural to apply it to the estimation of the most likely point of a global extremum, when function $f(x)$ is considered as a sample of some stochastic function.. The method developed by Strongin (1978) maximizes the likelihood of a parameter α of the stochastic function.

$$f(x) = \mu_x + \xi_x, \ x \in \{x_i\} \subset A \subset R, \ x_{i+1} > x_i, \ i = 1, \dots, n \qquad (5.4.1)$$

where ξ_x is a Wiener process with parameter cm, and μ_x is a fixed function which depends on paramter α

$$\mu_{x_i} = \begin{cases} -m(x_i - x_{i-1}), & i < i_\alpha \\ m(x_s + x_{s-1} - 2\alpha), & i = s \\ m(x_i - x_{i-1}), & i > i_\alpha \end{cases} \qquad (5.4.2)$$

Here c and m are both positive and index s is defined by the inequality

$$x_{s-1} \leq \alpha \leq x_s. \qquad (5.4.3)$$

In such a case the likelihood function is

$$H(z_n, \alpha) = -\sum_{i=1}^{n} \ln \left(cm \sqrt{2\pi(x_i - x_{i-1})} \right) - \frac{1}{2(cm)^2} \sum_{i=1}^{n} \frac{(y_i - y_{i-1} - \mu_{x_i})^2}{x_i - x_{i-1}}$$

$$(5.4.4)$$

Here $y_i = f(x_i)$, $i = 1, \dots, n$.

Part $h(z_n, \alpha)$ of the likelihood function (5.4.4) which depends on α can be expressed as

$$h(z_n, \alpha) = -\frac{2(\alpha - \alpha_s)^2}{c^2(x_s - x_{s-1})} + \frac{R_s}{mc^2}, \quad x_{s-1} \leq \alpha \leq x_s \tag{5.4.5}$$

where

$$z_n = (x_i, y_i, \ i = 1, \dots, n)$$

$$\alpha_s = \frac{x_s - x_{s-1}}{2} - \frac{y_s - y_{s-1}}{2m} \tag{5.4.6}$$

and

$$R_s = m(x_s - x_{s-1}) + \frac{(y_s - y_{s-1})^2}{m(x_s - x_{s-1})} - 2(y_s + y_{s-1}). \tag{5.4.7}$$

The necessary maximum condition is clearly $\alpha = \alpha_s$. Hence

$$h(z_n, \alpha_s) = \frac{R_s}{mc^2} .$$

This means that the maximum likelihood estimate of α is determined by (5.4.6) where the optimal index is obtained by condition

$$s = \arg \max_{1 \leq t \leq n} R_t . \tag{5.4.8}$$

It is easy to see that when the parameter $c \to 0$, the function $f(x)$ will converge probabilistically to the function μ_x. Hence the minimum $f(x)$ will also converge probabilistically to the minimum α of μ_x. The formal proof of this is given by Strongin (1978).

The unknown parameter m is estimated from m_n from the following expression

$$m_n = \begin{cases} 1, & M_n = 0 \\ \tau M_n, & M_n > 0 \end{cases} \tag{5.4.9}$$

where

$$M_n = \max_{1 \leq i \leq n} \left| \frac{y_i - y_{i-1}}{x_i - x_{i-1}} \right|, \quad r > 1 \tag{5.4.10}$$

The algorithm of Strongin (1978) is to make the $n+1$-th observation

$$x_{n+1} = \alpha_s$$

where α_s is from (5.4.6) and s is from (5.4.8).

It is supposed that $A = [a, b]$ and the first two observations $x_i, i = 1, 2$ are made at the ends of the interval A.

The stopping rule is

$$x_n - x_{n-1} \leq \varepsilon. \tag{5.4.11}$$

It was shown by Strongin (1978) that this algorithm can be regarded as asymptotically Bayesian if the losses are defined by a step function (3.5.1) and the variance parameter c converges to zero.

Unfortunately, it is difficult to generalize this simple algorithm to a multi-dimensional case, because then the m-dimensional probability distributions of the global minimum $x \in A \subset R^m$ must be calculated.

In the Bayesian case (5.1.1) we must calculate the one-dimensional probability distributions of $f(x) \in R$. The latter task is, of course, not an easy one, but it is obviously much simpler if $m > 1$.

The idea of Strongin's (1978) method is to transform m-dimensional space to one-dimensional space using Peano type mapping and then to perform the global optimization by a simple one-dimensional algorithm (5.4.6), (5.4.8).

5.5   The comparison of algorithms

The comparison of algorithms by computer simulation is a rather complicated task.

The first difficulty is how to select a good set of test problems.

The second one is how to define the quality of the algorithm.

The third difficulty is how to put all algorithms in equal conditions.

The convenient way is to select as test problems those problems which are published in well known papers and preferably have some practical connections. Such selection helps to arrange the 'competition' of algorithms without the authors being present.

Under the assumptions of this book the natural way is to define the quality of the algorithm as the relation between the average deviation of the global minimum and the computing time. Unfortunately, the computing time depends not only on the algorithm, but also on the computer implementation, including the parameters of the computer and the quality of the programming. These factors are eliminated if the number of function observations is substituted for the computing time.

The conditions for the algorithms can be made more consistent if we compare only algorithms which do not allow adaptation of parameters. The reason is that if the user is allowed to adjust the parameters of the algorithm to the given function, then we compare not only the quality of the algorithm, but also the abilities of the users. This means that during the competition the parameters of the competing algorithms should be defined automatically, without the user's help. The user's help is obviously desirable when real practical problems are considered.

The application of sophisticated stopping rules can also make the comparison of the algorithms more difficult, because the reason for good results will not be clear. Is it the good search procedure or the successful choice of the stopping rule? Thus, it would be desirable to compare the algorithms using the same simple stopping criterion. For example, to fix the number of observations. The other idea is to fix the accuracy level to be the same for all algorithms of optimization and to compare the average number of observations, see Dixon and Szego (1978). Obviously, in this case, the question as to whether the success or failure of the algorithm of optimization is due to the search procedure or to the stopping rule, will remain unanswered.

A good example of a set of test functions is the family of two-dimensional functions with parameters $a_{ij}, b_{ij}, c_{ij}, d_{ij} \in (0, 1)$.

$$f(x) = ((\sum_{i,j=1}^{I} \left(a_{ij} \sin (\pi i x^{(1)}) \ \sin (\pi i x^{(2)}) + b_{ij} \cos (\pi i x^{(1)}) \ \cos (\pi i x^{(2)})\right)^2$$

$$+ \sum_{i,j=1}^{I} (c_{ij} \sin (\pi i x^{(1)}) \ \sin (\pi i x^{(2)}) - d_{ij} \cos (\pi i x^{(1)}) \ \cos (\pi i x^{(2)})))^2)^{1/2}$$

(5.5.1)

where the number of components $I = 7$.

This family of functions satisfies our conditions, because it represents the stress function in an elastic square plate under a cross sectional load and is widely known, at least by the Soviet scientists. It was considered by Grishagin (1978) to test the different versions of the method of maximum likelihood (see Strongin (1978)). The full account of the conditions of the experiment was published, so it was relatively easy to arrange a sort of 'national competition' of algorithms by comparison of the maximum likelihood methods with other methods, widely known and used in the Soviet Union, such as LP type uniform search (Sobolj (1969)), the two versions of the one-step Bayesian method (Mockus (1972)) and the uniform random search (Monte-Carlo method).

In all cases, after the termination of the global search, a local optimization was performed using the simplex algorithm of Nelder and Mead (see Himmelblau (1972)). The local search was carried out only once from the best point of the global search. Considereing (5.5.1) it was noticed that to do the local optimization more than once is usually too expensive if the derivative cannot be calculated directly and so must be estimated using the function differences.

In addition to the completely automatic search, the purely interactive optimization performed by a well known expert in the field of global optimization was included.

Fifty sample paths corresponding to the random uniformly distributed parameters $a_{ij}, b_{ij}, c_{ij}, d_{ij} \in (0, 1)$ were considered.

The relation between the percentage of successful cases (when the global minimum was found) and the total number of observations $N_t = N + N_L$ (where N_L is the number of observations for the local Nelder-Mead search) is represented in Table 5.5.1 and Figure 5.5.1.

Index Iteration number	1	2	3	4	5	6
60						48
80	46	46		30	26	
90	60					
100		56		38		
105		62	56		44	
110						81
125	80	72				
135						92
140	88	86	68	44	68	
200			82	52		
240	96				84	
340					92	
370			94			
400			100	78	94	

Table 5.5.1.

The relation between the percentage of successful cases and the total number of observations

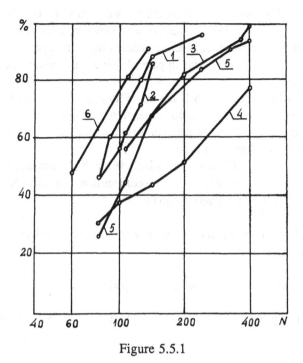

Figure 5.5.1

Relation of percentage of successful cases and the total number of observations for
different methods

Index 1 corresponds to the adaptive Bayesian algorithm (5.2.7).

Index 2 corresponds to the standard one-step Bayesian algorithm (5.1.1) with
Gaussian *a priori* distributions (4.4.2).

Index 3 corresponds to Strongin's algorithm (1978).

Index 4 corresponds to uniform random search (Monte-Carlo).

Index 5 corresponds to uniform deterministic LP-search.

Index 6 corresponds to the interactive search procedure performed by an
expert, see Shaltenis (1979).

The results of simulations of the adaptive Bayesian algorithm (5.2.7) using the
family of test function (5.5.1) were even better than those of the standard one-step
Bayesian methods. However, the results of simulation using another set of functions
considered by Mockus, Tieshis and Zilinskas (1978) were different. The reasons are
not yet clear. We only conclude that the adaptive Bayesian model can sometimes
better represent the real situation.

The sort of average deviation (the percentage of successful cases) which was
used by Grishagin (1978) is fairly natural for the methods of maximum likelihood

(see Strongin (1978)), because in those methods the zero-one type loss function is assumed. However, Grishagin's definition of the average deviation contradicts the assumption that the loss function is linear under which the Bayesian method (5.2.1) was developed. This difference can apparently give some advantage to maximum likelihood methods.

The family of functions (5.5.1) has some limitations as a set of test functions. It cannot be generalized to the multi-dimensional case without loss of physical meaning. It was published only in Russian, so is not easily available to most western scientists interested in global optimization.

A different family of functions was included in the 'international competition' of algorithms of global optimization. The 'competition' was arranged (without the presence of authors) by Dixon and Szego (1978).

Table 5.5.2 shows the average number of function evaluations required to locate the global minimum.

The methods were divided into three categories.

1) Trajectory type : Gomulka's implementation (1978) of Branin's method (1972).

2) Clustering type : Törn's method (1978), the two versions of Gomulka's implementation (1978) of Törn's method, and Price's method (1978).

3) Sampling type : Fagiuoli's method (1978), De Biase and Frontini's method (1978) and the one-step Bayesian method (5.1.3).

The idea of the trajectory type method is to move along the anti-gradient line to the local minimum then possibly to the local maximum and so on.

Clustering type methods define the region of global minimum as a result of the analysis of clusters representing the local minima.

Sampling type methods are sampling the whole area A in order to prepare the next stage of search.

Two families of functions and two fixed functions were considered.

1) Sheckel's family of functions of four variables,
2) Hartman's family of functions of three and six variables,
3) Branin's function of two variables,
4) Golstein's and Price's function of two variables.

Test Functions

Methods	Shekel m=4			Hartman n=4		Branin	Golstein Price
	n=5	n=7	n=10	m=3	m=6	m=2	m=2
Trajectory Type							
1 Gomulka - Branin	5500	5020	4860				
Clustering Type							
2 Törn	3679	3606	3874	2584	3447	1558	2499
3 Gomulka-Törn 1	6654	6084	6144	6766	11125	1318	1495
4 Gomulka-Törn 2	7085	6684	7352	2400	7600	1800	2500
5 Price	3800	4900	4400				
Sampling Type							
6 Fagiuoli	2514	2519	2518	513	2916	1600	158
7 De Biase-Frontini	620	788	1160	732	807	597	378
8 Mockus	1174	1279	1209	513	1232	189	362

Table 5.5.2

Average number of function evaluations to locate the global minimum

Shekel's family of functions:

$$f(x) = -\sum_{i=1}^{n} \frac{1}{(x - a_i)^T (x - a_i) + c_i}$$ (5.5.2)

$$x = (x_1, \dots, x_m)^T$$

$$a_i = (a_{i1}, \dots, a_{im})^T$$

$$c_i > 0, \quad A = \{x : 0 \leq x_j \leq 10, j = 1, \dots, m\}, \quad m = 4, n = 5, 7, 10.$$

i	a_i				c_i	n
1	4.0	4.0	4.0	4.0	0.1	5
2	1.0	1.0	1.0	1.0	0.2	
3	8.0	8.0	8.0	8.0	0.2	
4	6.0	6.0	6.0	6.0	0.4	
5	3.0	7.0	3.0	7.0	0.4	
6	2.0	9.0	2.0	9.0	0.6	7
7	5.0	5.0	3.0	3.0	0.3	
8	8.0	1.0	8.0	1.0	0.7	10
9	6.0	2.0	6.0	2.0	0.5	
10	7.0	3.6	7.0	3.6	0.5	

Table 5.5.3

The parameters of Shekel's function, $m = 4$.

i	a_{ij}			c_i	p_{ij}		
1	3.0	10.0	30.0	1.0	0.3689	0.1170	0.2673
2	0.1	10.0	35.0	1.2	0.4699	0.4387	0.7470
3	3.0	10.0	30.0	3.0	0.1091	0.8732	0.5547
4	0.1	10.0	35.0	3.2	0.03815	0.5743	0.8828

Table 5.5.4

The parameters of Hartman's function, $m=3$

i	a_{ij}						c_i	p_{ij}					
1	10.0	3.0	17.0	3.5	1.7	8.0	1.0	0.1312	0.1696	0.5569	0.0124	0.8283	0.5886
2	0.05	10.0	17.0	0.1	8.0	14.0	1.2	0.2329	0.4135	0.8307	0.3756	0.1004	0.9991
3	3.0	3.5	1.7	10.0	17.0	18.0	3.0	0.2348	0.1451	0.3522	0.2883	0.3047	0.6650
4	17.0	8.0	0.05	10.0	0.1	14.0	3.2	0.4047	0.8828	0.8732	0.5743	0.1091	0.0381

Table 5.5.5

The parameters of Hartman's function, $m=6$

Hartman's family of functions:

$$f(x) = -\sum_{i=1}^{n} c_i \exp\left(-\sum_{j=1}^{n} a_{ij}(x_j - p_{ij})^2\right) \tag{5.5.3}$$

$$x = (x_1, \ldots, x_m)$$

$$p_i = (p_{i1}, \ldots, p_{im})$$

$$a_i = (a_{i1}, \ldots, a_{im})$$

$$A = \{x : 0 \le x_j \le 1, \ j = 1, \ldots, m\}$$

$$n = 4, \ m = 3, 6.$$

Here p_i is the approximate location of the i-th local minimum,

a_i is proportional to eigenvalues of the Hessian at the i-th local minimum,

$c_i > 0$ is the 'depth' of the i-th local minimum (assuming that the interference of different local minima is not strong).

Branin's function :

$$f(x_1, x_2) = a(x_2 - bx_1^2 + cx_1 - d)^2 + e(1 - f) \cos x_1 + e, \tag{5.5.4}$$

$$a = 1, \ b = 5.1(4\pi^2), \ c = 5/\pi, \ d = 6, \ e = 10, \ f = 1/(8\pi)$$

$$-5 \le x_1 \le 10, \ 0 \le x_2 \le 15.$$

There are three minima, all global.

Goldstein's and Price's function :

$$f(x_1, x_2) = [1 + (x_1 + x_2 + 1)^2 (19 - 14x_1 + 3x_1^2 - 14x_2 + 6x_1x_2 + 3x_2^2)]$$

$$\times [30 + (2x_1 - 3x_2)^2 (18 - 32x_1 + 12x_1^2 + 48x_2 - 36x_1x_2 + 27x_2^2)]$$

$$-2 \le x_i \le 2.$$

There are four local minima, The global one is at the point $(0, -1)$ with the value $f = 3$.

Branin's trajectory method corresponds to the movement along the gradient-anti-gradient line from the local minimum to the local maximum and so on until, hopefully, the global minimum is reached.

The clustering method of Törn can be described in four steps.

1. Choose initial points by the uniform random distribution.

2. Push the points some steps towards the local minima using a local optimizer.

3. Find clusters by using a clustering analysis technique. If a tolerance condition is met, stop.

4. Take a sample of points from each cluster. Return to step 2.

The local optimizer implemented by Törn's algorithm consists of two elements: random search and linear search. The random part defines the direction of search. The deterministic linear part controls the step size.

If in the linear search at least one of the symmetric points is better than the origin, then the search is made in the direction of the first better point. If the linear search is successful, the two new points are put on the same line with double distance. If neither of the new points show improvement, then a new pair of observations is taken on a new random line with the distance divided by 2.

The local search stops if the step length is reduced to a given level. Later this level will be the starting step length when the local search is repeated.

The 'seed' points in the clustering process are selected giving priority to the points with the smallest function values. The cluster is grown by enlarging the hypersphere around the seed point as long as the density of cluster remains greater than k/v, where $v = \prod_{i=1}^{n} 4\lambda_i^{1/2}$ is the region spanned by the points $x_1, \ldots, x_k$. Here $\lambda_i, i = 1, \ldots, n$ ($\lambda_1 \leq \ldots \leq \lambda_n$) are the roots of the equation $\det (s - \lambda I) = 0$ where S is the 'covariance' matrix corresponding to the points $x_1, \ldots, x_k$.

One way to reduce the number of points in the cluster is to rank the points in each cluster in ascending order of function values and then choose every second, every third and so on in the hope that function values corresponding to points leading to a lower local minimum would also be smaller. This technique has been implemented by Törn (1979).

In the paper of Gomulka (1978) the two versions of Törn's algorithm were implemented. In the first version the usual Törn's algorithm for the local search was used. In the second version the variable metrics type algorithm, using numerical estimates of derivatives, was substituted for the random search routine for local optimization. The progress of the algorithm was controlled by the current gradient value in exactly the same way as the original implementation had been controlled by the step size.

The algorithm of Price (1978) combines random search and Nelder and Mead type routines into a single process. As an initial search, a predetermined number of trial points is chosen at random. The function is evaluated at each trial point. The positions and the function values corresponding to each point are stored. Then at each iteration a new trial point is selected. The function is evaluated at the new point and its value is compared with the worst point already stored. If the new point is better than the old, then the latter is replaced by the new. If not, then the trial is discarded and a fresh new point is chosen. To choose the new trial point the Nelder and Mead type method was combined with the random choice of simplexes from the old points.

The empirical rule for the size of the initial set is twenty-five points.

The stop criterion is not unique, e.g. after a specified number of function evaluations or when N points fall within a sufficiently small region.

The sampling method of Fagiuoli (1978) uses a so called stochastic automaton as a controller for the search adjusting the search probabilities according to past experiences.

In Fagiuoli's method the region of interest is covered by a set of 'cells' and a stochastic automaton is set up to govern the choice of cells. After a set of initial samples have been taken a cell is selected that is most likely to contain the global minimum. A local constrained minimum in the cell is obtained by the method of Biggs (1975). The stochastic automaton is then updated and the procedure repeated until the probability of further improvement is significantly reduced.

The algorithm of De Biase and Frontini is based on the sampling techniques proposed by Archetti (1975) which are based upon the idea of Chichinadze (1967). De Biase and Frontini constructed the spline type approximation to the probability distribution $F(y) = P\{f(x) < y\}$ of the objective function when the arguments x are distributed uniformly in the region A. An estimate is then made of the root $F(y) = 0$ which means the estimate of the global minimum. It helps to estimate the distance from the global minimum and so to arrange some stopping rules. The search is terminated using a local optimization algorithm starting from a fixed number of points selected from the best points. The important question of how many such points to choose remains open. This makes the comparison with the other methods more difficult.

The algorithm of the Bayesian sampling was of the standard one-step type (5.5.1) with *a priori* distribution (4.4.2). The number of observations, N, for the global search was fixed. The local minimization was repeated L times from the points $x_l, l = 1, \ldots, L$ with lowest deviations δ_l from the expected values. The deviation δ_l was determined as follows

$$\delta_l = \frac{f(x_l) - E\{f(x_l)|z_{N(l)}\}}{\sigma\{f(x_l)|z_{N(l)}\}} . \tag{5.5.5}$$

Here $z_{N(l)} = (f_i(x_i), x_i, i = 1, \ldots, N, i \neq l\}$ and $\sigma\{f(x_l)|z_{N(l)}\}$ means the conditional standard deviation. For the local minimization Tieshis (1975) version of the variable metrics algorithm was used. The local minimization was terminated if the norm of the gradient was less than $5 \cdot 10^{-4}$ or if the value of the function was decreasing less than $5 \cdot 10^{-4}$ in k iterations, where $k = 2$ if $n = 2$ and $k = 4$ if $n > 2$.

Table 5.5.2 shows the number of observations required to find the global minimum for different methods and functions. The numbers are from Dixon and Szego (1978).

The best performance was that of the method of De Biase and Frontini (1978). Unfortunately there is some ambiguity in the choice of starting points for the local search. The reasons for the success of the method also remain unclear. It is difficult to believe that the application of even the very good stopping rule can so drastically improve the performance of the Monte Carlo method, which is one of the least efficient methods, see for example Table 5.5.1, index 4.

The second best was the one-step Bayesian method. However this method is good only in the sense of the minimal number of observations which is a reasonable criterion only if one observation is very expensive. Otherwise the simpler methods such as, for example, Törn's (1978) or even Monte Carlo can be preferable. The simple Bayesian method (5.2.7) was not developed at the time of the 'competition'.

5.6 The Bayesian approach to global optimization with linear constraints

So far only rectangular feasible sets have been considered, because in this case the Bayesian risk function can be minimized approximately using the simplest methods of uniform search such as Monte Carlo or LP-search.

Now we shall consider a more general case when the feasible region A is defined by a system of linear inequalities

$$q_i^T x \leq b_i, \quad i = 1, \ldots, k. \tag{5.6.1}$$

We shall assume that A is bounded, nonempty and of full dimension, so that A is a polytope that contains interior points for which the inequalitites (5.6.1) are all satisfied as strict inequalitites.

It is reasonable to eliminate the linear equalities (if such are present) since this reduces the dimensionality of the problem correspondingly. This means that the case (5.6.1) can be considered as a fairly general one.

In case (5.6.1) the simple 'hit-and-run' algorithm can be used to optimize the Bayesian risk function, (5.1.1) or (5.2.7), because this algorithm generates a random sequence of interior points whose limiting distribution is uniform (see Mockus (1987)). The algorithm consists of the following steps.

Step 0:   Find an interior point x^0. Set $n = 0$.

Step 1:   Generate a direction vector v^n with equal probability from one of the m co-ordinate vectors and their negations.

Step 2:   Determine

$$\lambda_i = \frac{b_i - a_i^T x^n}{a_i^T v^n}, \; i = 1, \dots, k. \tag{5.6.2}$$

$$\lambda^+ = \min_{1 \le i \le k} \{\lambda_i \mid \lambda_i > 0\} \tag{5.6.3}$$

$$\lambda^- = \max_{1 \le i \le k} \{\lambda_i \mid \lambda_i < 0\} \tag{5.6.4}$$

Step 3:   Generate u from a uniform distribution on $[0, 1]$ and set

$$x^{n+1} = x^n + (\lambda^- + u(\lambda^+ - \lambda^-))v^n. \tag{5.6.5}$$

Step 4:   Set $n = n + 1$ and go to step 1 unless a stopping criterion $n = K$ is satisfied.

It was shown by Barbee et al (1986) that the limiting distribution of the sequence (5.6.5) is uniform on A.

The number K depends on how accurately we wish to perform the minimization of the risk function.

5.7 The Bayesian approach to global optimization with nonlinear constraints

Let us consider the case where the feasible region is determined by the following system of inequalities, at least one of them nonlinear

$$f_i(x) \leq 0, \quad (i = 1, \dots, k). \tag{5.7.1}$$

We shall assume, as in the linear case, that the set A defined by (5.7.1) is bounded, nonempty and contains interior points for which inequalities (5.7.1) are all satisfied and strict.

In such a case the 'hit-and-run' algorithm can be applied to optimize the Bayesian risk (5.2.7) (see Mockus (1987)). This is clearly reasonable if the calculation of the constraint functions $f_i(x)$ is at least $K \cdot M \cdot k$ times less expensive than the calculation of the objective function $f(x)$, where M is the number of constraint function $f_i(x)$ evaluations which are needed to find the intersection of the line $f_i(x) = 0$ and the corresponding co-ordinate vector v. If the intersection point can be calculated using an explicit expression, which is possible in the case when $f_i(x)$ is, for example, a low degree polynomial, then M means the fraction of the amount of calculations to define the intersection $f_i(x)$ and v in relation to the amount of calculations to determine the value of the objective function $f(x)$.

In the case of a nonlinear constraint function only step 2 of the hit-and-run algorithm is different:

Step 2. Determine

$$\lambda_i \text{ is such } \lambda : f_i(x^n + \lambda_i v^n) = 0, \tag{5.7.2}$$

$$\lambda_i^+ = \min_{1 \leq i \leq k} \{\lambda_i \mid \lambda_i > 0\} \tag{5.7.3}$$

$$\lambda_i^- = \max_{1 \leq i \leq k} \{\lambda_i \mid \lambda_i < 0\} \tag{5.7.4}$$

If the calculation of the intersections (5.7.2) is too complicated, for example if the equations (5.7.2) have many roots, then an alternative is the penalty function method when we pay some extra 'price' for the violation of constraints and this extra price is added to the value of the function to be optimized, which can be the objective function $f(x)$ or the risk function (5.2.7).

However, there are some serious problems related to the penalty function approach to global optimization. One of them is how to define the rectangular set B which includes a feasible set A.

The area of global search is supposed to be the rectangular one. The efficiency of methods of global search is inverse to the volume of the area of search, so the size of the set B should be as small as possible.

It can happen that the relation $\mu(A)/\mu(B)$, where μ is the Lebesgue measure, is so small that almost all observations will fall outside the feasible region of A so the problem of global search will degenerate to the search for a feasible region A, which can be performed more efficiently by some local methods. The simple example is

$$A = \{x : \sum_{i=1}^{n} x_i \leq 1, \; 0 \leq x_i \leq 1, \; i = 1, \dots, n\}.$$

In cases with very small relation $\mu(A)/\mu(B)$ the hit-and-run algorithm can be the reasonable alternative even if the equations (5.7.2) are complicated.

The function $f(x)$ is not always defined outside the area A. In this case some extension of $f(x)$ to the wider area B is needed, if we wish to apply the penalty function approach to minimize $f(x)$.

The correct definition of the penalty function is also not an easy task, because if the penalty function is too steep, then overflow can be the trouble. If the penalty function is not steep enough, then the minimising point can be far away from the feasible region A. This difficulty is not so important when the penalty function is applied in the case of local optimization because here the sequential procedure can be used, gradually increasing the steepness of the penalty function.

5.8   The Bayesian approach to global multi-objective optimization

Probably the most important idea in multi-objective optimization is the concept of Pareto optimality.

The point x_p is called Pareto optimal if there are no such points $x \in A$ that

$$f_i(x) \leq f_i(x_p), \; \text{for all } i = 1, \dots, K \tag{5.8.1}$$

$$f_i(x) < f_i(x_p), \; \text{for at least one } i \; . \tag{5.8.2}$$

We can approximately define the Pareto optimal set X as a by-product of any global optimization method which satisfies the condition of asymptotic density (4.2.65). Examples are the method of LP-search and the Bayesian method (5.2.7), which minimize the weighted sum of objectives $f_i(x)$. In both cases the approximation set X_n will converge to the Pareto optimal set X for any continuous functions $f_i(x)$ and compact feasible set A, in the sense that the distance r_n between any point in X and the nearest point in X_n will converge to zero (it follows from condition (4.2.65) that $r_n \to 0$ when $n \to \infty$).

The approximation set X_n is defined by the following conditions: the point $x_p \in X_n$ if there are no such points $x_j, j = 1, \dots , n$ that

$$f_i(x_j) \le f_i(x_p), \text{ for all } i = 1, \dots , K$$

$$f_i(x_j) < f_i(x_p), \text{ for at least one } i .$$

So if nothing is known about the relations and order of importance of the objective functions $f_i(x)$, then both methods (LP-search and Bayesian) can be regarded as asymptotically equivalent approximations of the Pareto set X. In this case the LP-search has the advantage of simplicity.

In practical problems the decision makers usually have some ideas about the relative importance of different components $f_i(x)$ and can express those ideas as some weights c_i corresponding to each f_i. In this case the Bayesian method minimising the weighted sum (see Mockus (1987))

$$h_c(x) = \sum_{i=1}^{K} c_i f_i(x), \ c_i > 0 \tag{5.8.3}$$

makes a better approximation of the Pareto set in areas where the scalarized objective is good comparing it with the uniform LP-search proposed by Sobolj and Statnikov (1981).

The reason is that in the Bayesian case the density of observations around the best values of the scalarized objective h_c is considerably greater than in the other areas. In the uniform LP-search case the density of observations is roughly the same in all the areas including the area around the largest possible value of the weighted sum h_c of objectives $f_i(x)$.

Let us present formally the advantage of the Bayesian approach to multi-objective optimization.

Let

$$r_0(c) = \max_{x \in S_0} \|x - x_0\| \tag{5.8.4}$$

Here S_0 is a set S_i (see (4.2.110) to (4.2.112)) around the point x_0 with the minimal observed value $h_c(x)$ obtained by the Bayesian method (5.2.7)

$$r_u(c) = \max_{x \in S_u} \|x - x_u\| \tag{5.8.5}$$

Here S_u is a set S_i around the point x_u with a minimal observed value of $h_c(x)$ corresponding to the uniform search.

In the case of constant μ_x^i and increasing σ_x^i as a function of $\|x - x_i\|$ (inside the sets S_i)

$$r_0(c) = r_0 \tag{5.8.6}$$

and

$$r_u(c) = r_u \tag{5.8.7}$$

Here

$$r_0 = \|x_0' - x_0\|, \quad x_0' \in \arg \min_{x \in S_0} \phi(\mu, \sigma, c)$$

and

$$r_u = \|x_u' - x_u\|, \quad x_u' \in \arg \min_{x \in S_u} \phi(\mu, \sigma, c)$$

where ϕ is from (4.2.79) and (5.1.1).

It follows from the definition of uniformity that in the strictly uniform case

$$r_a = \|x_a' - x_0\|$$

$$r_u = r_a$$

$$x_a' \in \arg \min_{x \in S_a} \phi(\mu, \sigma, c) \tag{5.8.8}$$

where S_a is a set S_i around the observation nearest to the mean value f_a of the function $f(x)$.

It is convenient to define the relative density of Bayesian and uniform observations around the point with the best observed value of $h_c(x)$ as the relation

$$K_a = \frac{r_u(c)}{r_0(c)} \tag{5.8.9}$$

From (5.8.6) to (5.8.8)

$$K_a = r_a/r_0. \tag{5.8.10}$$

Asymptotically from (4.2.120)

$$K = \lim_{a \to \infty} K_a = \left(\frac{f_a - f_0 + \varepsilon}{\varepsilon}\right)^{1/2}, \varepsilon > 0. \tag{5.8.11}$$

The asymptotic expression (5.8.11) defines the relative density of a Bayesian search for the Pareto optimal point near the global minimum of a scalarized objective $h_c(x)$.

In this case the correction parameter ε can be used to balance the subjective part of the information expressed by the scalarized objective $h_c(x)$ and the objective part of the information represented by the vector $f(x) = f_1(x), \ldots, f_k(x))$. If ε is large then we almost neglect the influence of the scalarized function $h_c(x)$ and simply seek the Pareto optimal point by an almost uniform search. If ε is small then almost all the observations are made near the minimum of the scalar $h_c(x)$ and only a small proportion of them will fall into the other parts of the feasible set A.

In some applications the multi-objective optimization is reduced to the constrained optimization. It is important to notice that such reduction is not always reasonable because it is much more convenient to consider the scalarized optimization (5.8.3) than the optimization with constraints, see section 5.7

$$\min_{x \in A} f_1(x)$$

$$f_i(x) \leq 0, \quad i = 2, \ldots, K.$$

In addition the solution of the constrained optimization problem does not necessarily belong to the Pareto optimum X.

5.9   Interactive procedures and the Bayesian approach to global optimization

The Bayesian approach to global optimization can be regarded as some formalized interactive optimization procedure. It was shown by Shaltenis (1979) that the actual interactive procedures performed by the experts are very similar in the average sense to the procedures of Bayesian search. It suggests that the formal (Bayesian) and informal (interactive) procedures of search can be united in the natural way: some of the observations may be made by interactive procedure, while other observations can be carried out in accordance with the Bayesian techniques. If, for example, we have some knowledge about the behaviour of a function, we can use it to fix the co-ordinates of the initial points. If we can get some ideas about the properties of the objective function, such as the number or location of local minima, then it is reasonable to do the additional interactive optimization after the Bayesian search. The latter possibility is especially important if the Bayesian procedure had reached its limit of observations.

The second important role of interaction is the choice of the parameters of the local search such as initial points and accuracy.

It can be desirable to control the accuracy and, correspondingly, the time of the minimization of the risk function interactively, by changing the number of its evaluations.

In the case of nonlinearly constrained global optimization, the interactive procedures may be necessary to make the decision as to whether the penalty function should be used or whether it is better to include the nonlinear constraints into the minimization of the risk function, see section 5.8.

The interactive procedures are almost inevitable in multi-objective optimization, for example to change the parameters c_i, during the process of optimization, to perform additional observations and to make other reasonable modification of the procedures of search during the optimization, see Mockus (1987).

It is very difficult to design formal procedures to adjust algorithms of the analysis of structures such as ANAL1 and the algorithms of optimization. The interactive procedures seem to be the most convenient way of arranging such adjustment.

5.10   The reduction of multi-dimensional data

The efficiency of the interactive procedures, which were discussed briefly in section 5.9, depends on the reduction of the multi-dimensional data such as the co-ordinates x of the points of observations and the results $f(x)$ to the two-dimensional display screen. The different methods of doing it can be approximately divided into two categories.

The first, 'mathematical' one, is to represent each point from multi-dimensional space in two dimensions with as small a distortion of the distances between the points as possible. The methods of representation, with different complexity and different accuracy, are discussed by Everitt (1978).

The second way can be called the 'psychological' one because there the multi-dimensional data is represented as some familiar image on the screen. Each parameter of the image is controlled by one dimensional data, for example by the value of one variable. The most natrural example of the familiar image is the human face, the different features of which, such as the width and height of the eyes, lips, nose, ears, head etc. are controlled by different variables. Because of the well known ability of human beings to recognize faces, such representation of the multi-dimensional points can be fairly efficient after some learning experience. The clear advantage of the representation of different multi-dimensional points as corresponding human faces is the familiarity of the two-dimensional image.

The disadvantage is the lack of symmetry, because the different variables are represented as different parts of the face which is only a partially symmetric figure. If we do not like the asymmetry of such representation more symmetric objects can be designed to represent multi-dimensional points. Naturally in the case of these objects we shall not have the familiarity of the human face, so probably more learning will be needed to relate the multi-dimensional points to the objects.

The usual diagrams which represent each variable x^i and the function $f(x)$ as corresponding columns can also be useful with a proper choice of scales.

The most natural and convenient way is to show on the computer display, if the system can be represented graphically, how the picture of the system is changing during the process of optimization. The objective function in such cases can be shown as one or more columns.

5.11   The stopping rules

A stopping rule is a natural part of any optimization algorithm. However, the importance of stopping rules depends on the type of problem under consideration. Usually the maximal number of iterations is limited by the available computer time. So the main task of a stopping rule is to stop the optimization when its continuation is obviously unreasonable. Such a moment can be clearly defined for example in convex problems because here we can usually obtain a good estimate of deviation from the optimal point. In the case of problems with known Lipshitz constant the estimate of distance from the global minimum is also obtainable and reasonably good. In the case of the family of continuous functions even the statistical estimation of the global minimum is possible only under some additional assumptions about the behaviour of the distribution function $F(y) = P(f(x) < y)$ on its left 'tail'. This means that the reliability of estimation depends completely on those assumptions which are not easy to verify even if verification is possible.

For example De Biase and Frontini (1978) and De Biase (1976) assume that $F(y)$ can be represented by spline functions of odd degree. Betro (1981) assumes some *a priori* distribution on a set of functions $F(y)$ which is updated as a result of a sample. In this way an *a posteriori* distribution is obtained for $F(y)$ for every y. Boender and Rinnooy Kan (1983) assume some *a priori* distribution on a set of local minima, namely that *a priori* each number L of local minima between 1 and ∞ is equally probable and that the relative sizes of regions of 'attraction' follow a uniform distribution on the $(L-1)$-dimensional simplex.

Zigliavski (1985) estimates a global minimum by the well-known, see Gnedenko (1943), asymptotic expression of extremal statistics. So the problem of distribution function can be reduced to a simpler problem of estimation of a single parameter α which shows the asymptotic rate of $F(y)$ in the vicinity of the global minimum. The parameter α can be estimated statistically from the sample. It can also be estimated using some *a priori* knowledge about the behaviour of $F(y)$ around the global minimum, assuming that in this region $F(y)$ can be represented as some homogeneous function of degree b. In both cases the theoretical framework is acceptable only if at least some of the observations are in the area of global minimum. However, in this case the usual stopping rules of local optimization can also be used. If no sample point is near the global minimum then the exact meaning of the 'estimate' of paramater α is not clearly defined, and so the corresponding estimate of the global minimum remains without proper justification. The use of an asymptotic expression for $F(y)$ in the case of a finite number of observations also needs some additional explanation.

Zilinskas (1986) estimates empirically the number of local minima. Optimization is stopped if the estimated number of local minima is equal to some fixed number L, $0 < L \leq 20$; see section 9.10 (the description of algorithm UNT). The second stopping condition of Zilinkas is M-multiply descent to the same local minimum (in the case of UNT the number M is 3).

Mockus (1963) and (1967) assumed lognormal $F(y)$ and estimated the minimum by the method of maximum likelihood.

A brief survey of different assumptions used in the estimation of global optimum is given by Zigliavski (1985).

Different authors support their assumptions by corresponding computer simulation. In all reported cases considerable improvement in the efficiency of optimization was noticed. This means that for each assumption there exists a set of functions for which the corresponding stopping rule works well.

Unfortunately we must not be too optimistic because here we deal with the problem of statistical extrapolation, namely the statistical estimation of the global minimum which, by definition, is usually outside the area covered by the sample. This means that the quality of estimation can depend more on assumptions than on the sample data. It is the natural price to pay for extrapolation outside the sample region when there is no certainty that the behaviour of $F(y)$ will not change completely outside the sampled area. The problem is made even more difficult by the high sensitivity of estimates of global minimum to the unknown parameters which define the shape of the 'tail' of the distribution function $F(y)$.

CHAPTER 6

THE ANALYSIS OF STRUCTURE AND THE SIMPLIFICATION OF THE OPTIMIZATION PROBLEMS

6.1 Introduction

It is well known that the complexity of optimization problems (in the sense of the difficulty of the solution) usually increases much faster than linearly with the number of variables. This means that it is easier to solve, for example, a sequence of n one-dimensional problems than one n-dimensional problem of optimization. The difference is very great in multi-modal problems. It is usually impossible, or almost impossible, to find the exact solution of multi-modal optimization. It is permissible for the solution of the simplified problem to be not exactly the same as the solution of the original one, if the difference does not exceed the error of the optimization. In the case of the Bayesian approach it is natural to consider the average error of simplification.

The most obvious way of simplifying the optimization problem is to fix the values of some of the less 'influential' variables and to optimise the remaining problem with the reduced number of variables, if the expected error of such simplification does not exceed the permissible level.

The other way is to replace the n-dimensional problem by a sequence of one-dimensional problems if the estimated influence of pairs of variables is not significant. Many different ways of simplifying the optimization problem exist, but it is important to estimate the expected error caused by the simplification.

It was shown by Shaltenis and Radvilavichute (1977) that under some assumptions there exists a relation between the expected error of simplification and the 'structural characteristics' which can be estimated using results of observations.

One of the interesting applications of structural characteristics is the optimal enumeration of variables in the case of LP-search. It was shown by Shaltenis (1982) that the efficiency of LP-search can be significantly increased if the enumeration of variables corresponds to the highest structural characteristics D_i.

6.2 Structural characteristics and the optimization problem

We shall consider the following orthogonal expansion of the function $f(x)$ defined on the unit hypercube A

$$f(x) = c + \sum_{i=1}^{n} f_i(x_i) + \ldots +$$

$$+ \sum_{1 \le i_1 < \ldots < i_s \le n} \ldots \sum f_{i_1 \ldots i_s}(x_{i_1}, \ldots, x_{i_s}) + \ldots + f_{1 \ldots n}(x_1, \ldots, x_n) \qquad (6.2.1)$$

where c is the mean value of the function $f(x)$ on the cube A under the assumption of the uniform distribution of $x \in A$, $s = 2, \ldots, n-1$

$$c = \int_A f(x_1, \ldots, x_n) \, dx_1 \ldots dx_n. \qquad (6.2.2)$$

The expansion (6.2.1) is unique if we demand that the mean values of the components of the expansion should be equal to zero (with the exception, of course, of the component c).

The expansion (6.2.1) exists for a wide class of functions which are continuous at any point $x \in A$ with the exception of dyadic rational points, see Shaltenis and Varnaite (1976).

The structural characteristics of the variables and their groups are defined by Shaltenis and Varnaite (1976) as the variances of corresponding components of the expansion (6.2.1), namely

$$D_{i_1 \ldots i_s} = \int_A [f_{i_1 \ldots i_s}(x_{i_1}, \ldots, x_{i_s})]^2 \, dx_{i_1} \ldots dx_{i_s}. \qquad (6.2.3)$$

It is convenient to normalise the structural characteristics by the following condition

$$\sum_{1 \le i_1 < \ldots < i_s \le n} \ldots \sum D_{i_1 \ldots i_s} = 1, \quad s = 1, \ldots, n. \qquad (6.2.4)$$

The variance of $f(x)$

$$D = \int_A (f(x_1, \ldots, x_n) - c)^2 \, dx_1 \ldots dx_n = \sum_{i=1}^{n} D_i$$

$$+ \sum_{1 \le i_1 < i_2 \le n} \sum D_{i_1 i_2} + \ldots + D_{12 \ldots n} \qquad (6.2.5)$$

because the components of the expansion are orthogonal.

It was shown by Shaltenis and Radvilavichute (1977) that under the independence conditions the mean error of simplification is proportional to the sum of the structural characteristics of the variables or their groups, which are eliminated from the process of optimization, for example, by fixing their values. This means that to minimise the error of simplification we should optimise the variables with the largest structural characteristics.

6.3 The estimation of structural characteristics

The algorithm of estimation of the structural characteristics $D_{i_1 \ldots i_s}$ is based on the Fourier-Haar expansion of the components of the expression (6.2.1)

$$f_{i_1 \ldots i_s}(x_{i_1}, \ldots, x_{i_s}) = \sum_{k_1, \ldots, k_s = 1}^{l} c_{k_1 \ldots k_s}^{i_1 \ldots i_s} h_{k_1}(x_{i_1}) \ldots h_{k_s}(x_{i_s}) \qquad (6.3.1)$$

where $c_{k_1 \ldots k_s}^{i_1 \ldots i_s}$ are the Fourier-Haar coefficients.

The Haar functions are the step functions

$$h_k(x) = \begin{cases} 2^{(m-1)/2}, & \text{if } \dfrac{j-1}{2^{m-1}} \le x \le \dfrac{j-1/2}{2^{m-1}} \\[2ex] -2^{(m-1)/2}, & \text{if } \dfrac{j-1/2}{2^{m-1}} \le x < \dfrac{j}{2^{m-1}} \\[2ex] 0, & \text{in the other cases} \end{cases} \qquad (6.3.2)$$

where the integers m and j are defined by the conditions

$$k = 2^{m-1} + j, \quad m = 1, 2, \ldots \quad j = 1, \ldots, 2^{m-1}$$

and $h_1(x) = 1$.

From (6.2.1) and (6.3.1) it follows that

$$f(x) = \sum_{k_1, \ldots, k_n = 1}^{l} [\sum_{i=1}^{n} C_{k_i}^i H_{k_i}(x_i) + \ldots +$$

$$+ \sum_{1 \le i_1 < \ldots < i_s \le n} \ldots \sum C_{k_1 \ldots k_s}^{i_1 \ldots i_s} H_{k_1}(x_{i_1}) \ldots H_{k_s}(x_{i_s}) + \ldots +$$

$$+ C_{k_1 \ldots k_n}^{12 \ldots n} H_{k_1}(x_1) \ldots H_{k_n}(x_n)]. \tag{6.3.3}$$

Here the sum of the indices $k_1, \ldots, k_n$ has l^n components. The functions $H_k(x)$, $k = 2, \ldots, l$ are the Haar functions, normalised by the following conditions

$$H_{k_i}(x) = \frac{h_{k_i}(x)}{2^{(m-1)/2}} \tag{6.3.4}$$

and $C_{k_1 \ldots k_s}^{i_1 \ldots i_s}$ are the Fourier-Haar coefficients modified to satisfy (6.3.4), (6.3.3) and $H_1(x) = 1$.

The following algorithm of the estimation of structural characteristics considers not all but only the largest components of the Fourier-Haar expansion.

Let us begin with the estimation of the coefficient $C_{k_i}^i$, $k_i = 2, \ldots, l$; $i = 1, \ldots, n$. The values of $C_{k_i}^i$ are estimated by the average values of $f(x)$, and the structural characteristics of the corresponding components are estimated by the mean square of $f(x)$ in the interval where $H_{k_i}(x)$ should not be zero in accordance with condition (6.3.2). The estimate is denoted by D_{H_k}. The largest D_{H_k} is defined and the estimated value of the corresponding component of the Fourier-Haar expansion is subtracted from $f(x)$ and so the modified function $f_1(x)$ is calculated.

The same process is repeated with the sequence of modified fuctions $f_i(x)$, $i = 1, \ldots, s$. In such a way a fixed number of the largest structural characteristics of the Fourier-Haar expansion are defined.

Because of the orthogonality of the Fourier-Haar expansion the structural characteristics of variables D_i, $i = 1, \ldots, n$ can be defined simply by summing the structural characteristics of the corresponding Fourier-Haar functions in accordance with the expression (6.3.1).

A similar algorithm is used to define the structural characteristics of pairs of variables as the sums of the structural characteristics of the two-dimensional components of the Fourier-Haar expansion.

The estimation of the structural characteristics of a group of more than two variables can also be done in similar way, but it is much more complicated computationally.

6.4   The estimation of a simplification error

We shall suppose that the vector of variables x is decomposed into two parts, x_1 and x_2, with the number of components $m < n$ and $n - m$, respectively. We shall consider the simplification when

$$x^0 = (x_1^0, x_2^*). \tag{6.4.1}$$

Here $x_1^0 \in \arg \min_{x_1} f(x_1, x_2^*)$ and x_2^* is fixed.

It was suggested by Shaltenis et al (1976), (1977) that the average error of simplification (6.4.1) can be estimated by the structural characteristics

$$\delta = D_2 \tag{6.4.2}$$

or by the relation

$$k_1 = \frac{D_{12}}{D_1 + D_{12}} \tag{6.4.3}$$

In the more general case (where not only two but more different groups of variables are considered) the average error of simplification is estimated by Shaltenis (1976) as

$$k_i = \frac{D_{i\bar{i}}}{D_i + D_{i\bar{i}}} \tag{6.4.4}$$

Here $\bar{i}$ corresponds to all indices with the exception of the index i.

6.5   Examples of the estimates

Table 6.5.1 shows the estimates of the simplification error of different optimization problems. The first row corresponds to the Steiner problem, see section 8.14, with four variables. The second row shows the coefficients k_i from (6.4.4) of the electricity meter problem, see section 8.2, in the case where all except two variables are fixed. The third row corresponds to the very simple test problem $f(x) = x_1 x_2$.

No.	Problem	n	k_1	k_2	k_3	k_4
1	Steiner	4	0.65	0.65	0.84	0.84
2	Electricity meter	2	0.59	0.88		
3	$f(x) = x_1 x_2$	2	0.25	0.25		

Table 6.5.1

The simplification error of different optimization problems

Table 6.5.2 shows the structural characteristics of the Steiner problem.

$D_1 = 17\%$		$D_2 = 17\%$		$D_3 = 6\%$	$D_4 = 6\%$
$D_{12} = 7\%$	$D_{13} = 8\%$ $D_{14} = 8\%$	$D_{23} = 8\%$	$D_{24} = 8\%$	$D_{34} = 8\%$
	$D_{123} = 2\%$		$D_{124} = 2\%$		
		$D_{1234} = 8\%$			

Table 6.5.2

The structural characteristics of the Steiner problem

Table 6.5.3 shows the structural characteristics of variables and their groups of the well known test problem No. 11, see Himmelbau (1972):

Minimise the function of five variables

$$f(x) = 5.3578547\, x_3^2 + 0.8356891 x_1 \cdot x_5 + 37.293239 x_1 - 40792.141$$

under the conditions that

$$0 \leq 85.334407 + 0.0056858 x_2 x_5 + 0.0006262 x_1 x_4 - 0.0022053 x_3 x_5 \leq 92$$

$$90 \leq 80.51249 + 0.0071317 x_2 x_5 + 0.0029955 x_1 x_2 + 0.0021813 x_3^2 \leq 110$$

$$20 \leq 9.300961 + 0.0047026 x_3 x_5 + 0.0012547 x_1 x_3 + 0.0019085 x_3 x_4 \leq 25$$

$78 \leq x_1 \leq 102$
$33 < x_2 \leq 45$
$27 \leq x_3 \leq 45$
$27 \leq x_4 \leq 45$
$27 \leq x_5 \leq 45.$

D_5	D_3	D_2	D_{25}	D_1	D_{35}
0.20	0.20	0.19	0.07	0.05	0.03

Table 6.5.3

The structural characteristics of the Himmelbau problem No. 11

We can see that the sum of only six of the structural characteristics is 82%.

Table 6.5.4 shows the difference between the estimate of simplification error (6.4.2) and its actual value defined by numerical integration.

No.	Simplification	Average error	
		Estimated	Actual
1	Elimination of x_4	0.15	0.11
2	Elimination of x_1	0.17	0.14

Table 6.5.4

The difference between the estimated error of simplification and its actual value

One of the possible applications of the structural characteristics is the 'optimal' numeration of variables in the LP-search. The influence of the order of numeration of variables was investigated using the well known Mandelshtam problem, see Shaltenis (1982).

$$\min_{0 \leq x_k \leq 2\pi} \left(\max_{0 \leq \psi \leq 2\pi} |f(\psi)| \right)$$

where

$$f(\psi) = \sum_{k=1}^{n} u_k \cos(k\psi + x_k).$$

Table 6.5.5 shows the average efficiency ratio of LP-search depending on the numeration order.

The efficiency ratio is defined as the relation

$$K = N^*/N$$

where N and N^* satisfy the following condition

$$F(N) = F^*(N^*).$$

Here $F(N)$ and $F^*(N^*)$ show how the minimal value $F(N)$ of the function $f(x)$ which is obtained after N observations depends on the number of observations. The number N corresponds to the LP-search and the number N^* to the Monte Carlo search.

Numeration order	$K(\%)$
1 2 3 4	3.2
4 3 2 1	− 2.4

Table 6.5.5

The efficiency ratio of LP-search

The second row of the table shows that the efficiency of the LP-search can be less than that of the Monte Carlo search if the order of numeration is opposite to the order of decreasing structural characteristics.

CHAPTER 7

THE BAYESIAN APPROACH TO LOCAL OPTIMIZATION

7.1 Introduction

There are no practical reasons for using the Bayesian approach to optimize convex functions without noise. The well known methods of 'second order' based on a quadratic approximation such as variable metrics or conjugate gradients are apparently nearly optimal and usually ensure a superlinear convergence. However, it is only when there is no noise. The presence of even a small amount of noise can change the situation completely.

The methods of second order which are so good in the absence of noise are not at all popular in stochastic programming. Here the methods of practical use are mostly based on a linear approximation, hence they are of gradient type. An example of this can be the method of a stochastic quasigradient, see Ermolyev (1976).

These methods are simple and perform reasonably well. However, some problems remain which can be considered most naturally in the Bayesian framework. One of them is the problem of optimal step size.

The well known rules for controlling step size are derived from the convergence conditions, see Ermolyev (1976) and Blum (1954). At best they ensure only a sort of asymptotic optimality of a minimax type (see Wasan (1969)). So the problem of an optimal step size when the number of iterations is not large remains open and can be conveniently considered by the Bayesian approach, see Mockus (1984). The idea underlying the Bayesian approach is that the step size should minimize the conditional expectation along the line of the stochastic gradient. This idea was discussed in general terms by Urjasjev (1986) but the relation of the actual Urjasjev's algorithm to the general idea was not made totally clear.

7.2 The one-dimensional Bayesian model

It is convenient to begin with consideration of the one-dimensional case.

The statistical model will be defined as follows: Let

$$f''_x = d^2f(x, \omega)/dx^2, \; f'_x = df(x, \omega)/dx \;\; \text{and} f_x = f(x, \omega), \; x \in A, \; \omega \in \Omega.$$

Suppose that $x = 0$ is an initial point and that f''_x and f_x are stochastic functions defined by some *a priori* distribution such that

$$E\left\{\int f''_x dx\right\} = \int E\{f''_x\} dx \ ,$$

$$E\left\{\int f_x dx\right\} = \int E\{f'_x\} dx \qquad\qquad (7.2.1)$$

and there exists such positive a that

$$E\{f''_x\} = a. \qquad\qquad (7.2.2)$$

Then from (7.2.1) and (7.2.2)

$$E\{f'_x\} = ax + d \qquad\qquad (7.2.4)$$

and

$$E\{f_x\} = 1/2\, ax^2 + dx + b. \qquad\qquad (7.2.5)$$

Assume that the observation is made at the point $x = 0$.
From (7.2.2) and (7.2.4) it follows that

$$E\{f'_0\} = a, \qquad\qquad (7.2.7)$$

$$E\{f_0\} = d. \qquad\qquad (7.2.8)$$

The natural and convenient loss function is the linear one, see Theorem 3.2.1

$$l(x) = f_x - \inf_{x \in A} f_x. \qquad\qquad (7.2.10)$$

The corresponding risk function

$$r'_x = E\{f_x\} - E\{\inf_{x \in A} f_x\}. \qquad\qquad (7.2.11)$$

From (7.2.11) and (7.2.5) it follows that

$$r'_x = 1/2\, ax^2 + dx + b - E\{\inf_{x \in A} f_x\} \qquad\qquad (7.2.12)$$

or omitting $E\{\inf_{x \in A} f_x\}$

$$r_x = 1/2\, ax^2 + dx + b. \tag{7.2.13}$$

If parameter a is positive then the Bayesian decision x_1 can be defined by the following equation (when the initial observation is at the point $x = 0$)

$$ax + d = 0.$$

Hence the Bayesian decision

$$x_1 = -d/a, \text{ if } a > 0 \tag{7.2.14}$$

and

$$x_1 = \arg\min_{x \in A} (1/2\, ax^2 + dx), \text{ if } a \leq 0. \tag{7.2.14^a}$$

Later, for simplicity (to avoid the calculation of the intersection of the line of search with the border of area A which is not simple in the multi-dimensional case) the condition (7.2.14^a) will not be used. Instead we shall simply keep the estimate of parameter a positive; see expressions (7.3.44) and (7.3.45).

The unknown parameters d, a in (7.2.14) are not directly observable. Conditions (7.2.7) and (7.2.8) show that the derivatives f_0' and f_0'' can be used as estimates of the unknown parameters a and d, unbiased in the *a priori* probability space.

Expression (7.2.14) shows that the Newtonian method can be regarded as the one-step Bayesian method which is optimal in the sense of an average deviation under conditions (7.2.1) and (7.2.2). This conclusion remains true in the stochastic case as well, when we observe only the sums

$$h_x'' = f_x'' + g_x'' \tag{7.2.15}$$

$$h_x' = f_x' + g_x' \tag{7.2.16}$$

and

$$h_x = f_x + g_x. \tag{7.2.16.1}$$

Here g_x'', g_x' and g_x are the independent stochastic variables defined by the noise probability distributions P_0 with zero expectations and variances σ''^2, σ'^2, σ^2, respectively.

Conditions (7.2.15) and (7.2.16) along with (7.2.7) and (7.2.8) mean that h_0'', h_0' are unbiased estimates of a and d also in the more general probability space which is a product of *a priori* P and noise P_0 probability spaces.

By the use of the estimates h_0', h_0'' of d, a in (7.2.14) we have

$$x_1 = -h_0'/h_0'', \text{ if } h_0'' > 0. \tag{7.2.17}$$

In this book we are consistently following the assumption that only the values of the function h_x but not the values of the derivatives h_x', h_x'' can be observed directly. In such a case h_0' and h_0'' can be expressed only indirectly, for example, as finite differences h' and h'' respectively.

$$h' = (h_{q_0} - h_{-q_0})/(2q_0) \tag{7.2.18}$$

and

$$h'' = (h_q - 2h_0 + h_{-q})/q^2. \tag{7.2.19}$$

Suppose that

$$q^2 = q_0. \tag{7.2.20}$$

Then by the substitution of h', h'' for h_0', h_0'' from (7.2.17)

$$x_1 = -\Delta_0/2\alpha_0, \text{ if } \alpha_0 > 0 \tag{7.2.21}$$

where

$$\Delta_0 = h_{q_0} - h_{-q_0} = h'2q_0 \tag{7.2.22}$$

and

$$\alpha_0 = h_q - 2h_0 + h_{-q} = h''q^2. \tag{7.2.23}$$

Formula (7.2.21) defines the length of the next step of a local Bayesian algorithm, assuming that the next observation is the last one. Expression (7.2.21) gives the minimum of the risk function when parameters a and d are estimated by h'' > 0 and h' from (7.2.18) and (7.2.19), respectively.

Suppose that we can 'remember' only the current values of the first and second derivatives at the point x_n or their estimates.

Then according to the one-step Bayesian approach from (7.2.14) we have

$$x_{n+1} = x_n - 2k_n f'_{x_n}/f''_{x_n} \tag{7.2.24}$$

if the derivatives $f'_{x_n}, f''_{x_n} > 0$ are known, or

$$x_{n+1} = x_n - 2k_n \, h'_n/\, h''_n \tag{7.2.25}$$

if the estimates $h'_n, h''_n > 0$, where $h'_n = h'_{x_n}$ and $h''_n = h''_{x_n}$ (see (7.2.15) and (7.2.16)) are known, or

$$x_{n+1} = x_n - k_n \, \Delta_n/\alpha_n \tag{7.2.26}$$

if the estimates $h', h'' > 0$ from (7.2.18) and (7.2.19) are known. In (7.2.26) $\Delta_n = h'2q_0$ and $\alpha_n = h''q_0k_0 = 1/2$.

Here the index n of the parameter k_n shows that the step length may depend on n. The dependence will be defined more precisely later in section 7.3 from the convergence conditions.

The equalities (7.2.24) to (7.2.26) define the one-step Bayesian methods when the search is performed along the line, for example along the stochastic estimate of the gradient, without the comparison of functional values. It corresponds to the classical methods of stochastic approximation (see Ermolyev (1976) and Wasan (1969)).

In the Bayesian case it is reasonable to go from the current point to the next one if the expected gain is non-negative

$$x_{n+1} = \begin{cases} x_n - \beta_n, & \text{if } h_{x_n - \beta_n} < h_{x_n} + \varepsilon_n \\ \\ x_n & , \text{ if } h_{x_n - \beta_n} \geq h_{x_n} + \varepsilon_n \text{ and } l_n < L \end{cases} \tag{7.2.27}$$

where $\varepsilon_n > 0$.

From (7.2.24)

$$\beta_n = 2k_n f'_{x_n}/f''_{x_n}, \text{ if } f''_{x_n} > 0 \tag{7.2.28}$$

From (7.2.25)

$$\beta_n = 2k_n \, h'_n/\, h''_n, \text{ if } h''_n > 0 \tag{7.2.29}$$

and from (7.2.26)

$$\beta_n = k_n \Delta_n / \alpha_n, \text{ if } \alpha_n > 0. \tag{7.2.30}$$

The number l_n shows how many times the lower inequality of condition (7.2.27) has occurred.

Condition (7.2.27) means that we shall not move from the 'old' n-th observation as long as it remains better than the new one by at least ε_n, but not for more than L repetitions.

The restriction to no more than L repetitions is because we want to keep the asymptotic properties of the Bayesian methods similar to those of the usual methods of stochastic approximation. It allows us to use the same convergence conditions.

The reason why ε_n should be more than zero can be explained in the following way. Expression (7.2.27) shows that the current h_{x_n} is a minimum of several stochastic variables while $h_{x_n - \beta_n}$ is a stochastic variable. So if $\varepsilon_n = 0$, then the expected tendency is to remain at the current point. Equality (7.2.27) shows that this tendency will be balanced if the following symmetry condition holds:

$$E\{h_{x_n - \beta_n}\} = E\{h_{x_n}\} + \varepsilon_n. \tag{7.2.31}$$

It was shown by Senkiene (1983) that if $h_{x_n - \beta_n}$ can be regarded as a Gaussian random variable ξ_i with zero expectation and standard deviation σ and if h_{x_n} can be considered as a minimum of n independent Gaussian random variables with the same parameters, then $\varepsilon_1 = 0$, $\varepsilon_2 = \sigma / \sqrt{\pi}$.

From Senkiene (1986)

$$\varepsilon_n \to \sigma / \sqrt{(2\pi)}, \; n \to \infty.$$

From a Monte Carlo simulation, $n = 200$

$$\varepsilon_n = 0{,}9 \, \sigma / \sqrt{\pi}. \tag{7.2.32}$$

7.3   The convergence of the local Bayesian algorithm

Denote by $\rho(x_i, x_j)$ the average distance between the two points x_i and x_j

$$\rho^2(x_i, x_j) = E_0\{(x_i - x_j)^2\}, \tag{7.3.1}$$

where the expectation E_0 is defined on the noise probability space. If x_i and x_j are fixed, then

$$\rho^2(x_i, x_j) = (x_i - x_j)^2. \tag{7.3.2}$$

If the stochastic variables $x_i = x_i(\omega)$ and $x_j = x_j(\omega)$, $\omega \in \Omega$ are different only on a subset of Ω of zero probability P_0, then we shall regard them as equal:

$$x_i = x_j, \text{ if } x_i(\omega) = x_j(\omega) \pmod{P_0}. \tag{7.3.3}$$

In such a case average distance can be regarded as usual distance defining some metric space.

Let us denote transformations (7.2.24), (7.2.25), (7.2.26) and (7.2.27) as $T_1(x)$, $T_2(x)$, $T_3(x)$ and $T_4(x)$ respectively. Denote the distance between the transformations T_i and T_j as

$$\rho^2_{ij} = \rho^2(T_i(x_n), T_j(x_n)).$$

From (7.2.18) and (7.2.19)

$$h' = (h_{x_n + q_0} - h_{x_n - q_0})/(2q_0) = f'_{x_n} + \delta' + g' \tag{7.3.4}$$

where

$$\delta' = f' - f_{x_n} \tag{7.3.5}$$

$$f' = (f_{x_n + q_0} - f_{x_n - q_0})/(2q_0) \tag{7.3.6}$$

and

$$g' = (g_{x_n + q_0} - g_{x_n - q_0})/(2q_0). \tag{7.3.7}$$

Here g' is a random variable with zero expectation and standard deviation

$$\sigma' = \sigma/(\sqrt{(2l')}q_0) \tag{7.3.8}$$

where l' is the number of observations performed at each of the two points $x_n + q_0$ and $x_n - q_0$.

Similarly

$$h'' = f'_{x_n} + \delta'' + g'' \tag{7.3.9}$$

where

$$\delta'' = f' - f'_{x_n} \tag{7.3.10}$$

$$f' = (f_{x_n+q} - 2f_{x_n} + f_{x_n-q})/q^2 \tag{7.3.11}$$

and

$$g'' = (g_{x_n+q} - 2g_{x_n} + g_{x_n-q})/q^2. \tag{7.3.12}$$

Here g'' is a random variable with zero expectation and standard deviation

$$\sigma'' = \sqrt{2}\sigma/q^2 \left(1/l'' + 2/l\right)^{1/2} \tag{7.3.13}$$

where l'' is the number of observations performed at each of the points $x = x_n + q$ and $x = x_n - q$ and l is the number of observations at the point $x_n u$.
From (7.2.20)

$$\sigma'' = \sqrt{2}\sigma/q_0 \left(1/l'' + 2/l\right)^{1/2} \tag{7.3.14}$$

It follows from (7.2.20), (7.3.5), (7.3.6), (7.3.10) and (7.3.11) that deterministic errors of the approximations δ' and δ'' will converge to zero only if $q_0 \to 0$. Unfortunately in such a case the variances σ'^2 and σ''^2 of the corresponding stochastic errors of approximation will increase indefinitely in accordance with (7.3.8) and (7.3.14).

The square of the distance between the first and third transformations from (7.2.24) and (7.2.26)

$$\rho_{13}^2 = (2k_n)^2 E_0\left\{\left(\frac{f'_{x_n}}{f_{x_n}''} - \frac{h'}{h''}\right)^2\right\} \tag{7.3.15}$$

If there exists a Lipschitz constant $1 - \varepsilon$, $0 < \varepsilon \le 1$ of transformation T_1 and

$$\sup \rho_{13} = \rho < \infty$$

then from (7.2.41) and Collatz (1964) the expression of the upper bound of error of method (7.2.25) is as follows

$$\rho(x_n, x^*) \le \rho/\varepsilon \tag{7.3.16}$$

where x^* is the point of the minimum.

In the general case it is difficult to calculate ρ_{13} from (7.3.15). It is less complicated, if the error of the estimation of the second derivative is small. It follows from (7.3.10) and (7.3.14) that the error of the estimations of the second derivative will be small if $\delta'' \to 0$ and $l'' \to l \to \infty$.

LEMMA 7.3.1. *Suppose that f is twice differentiable and there exist integrable functions s_1, s_2 and a positive number β such that*

$$|g'| \leq s_1 \pmod{P_0}$$

$$\left(|f'_{x_n} + \delta'' - g''|\right)^{-1} \leq s_2 \pmod{P_0} \tag{7.3.17}$$

and

$$\inf f'_{x_n} = \beta > 0. \tag{7.3.18}$$

Then from $l'' \to \infty$ and $\delta'' \to 0$ it follows that in the case when the second derivative f'_{x_n} is known

$$\rho_{13}^2 \to (2k_n)^2 \left(f'^2_{x_n}\right)^{-1} (\delta'^2 + \sigma'^2) \tag{7.3.19}$$

and in the case when the estimate h'' of the second derivative is known

$$\rho_{13}^2 \to (2k_n)^2 (h''^2)^{-1} (\delta'^2 + \sigma'^2) \tag{7.3.20}$$

Proof. Let

$$\gamma^2 = \left(\frac{f'_{x_n}}{f''_{x_n}} - \frac{h'}{h''}\right)^2$$

From (7.3.4) and (7.3.9)

$$\gamma^2 = \left(\frac{f'_{x_n}}{f''_{x_n}} - \frac{f'_{x_n} + \delta' + g'}{f''_{x_n} + \delta'' + g''}\right)^2 \tag{7.3.21}$$

The expectation of g'' is zero and the variance converges to zero with probability P_0.

From this and from the assumption that $\delta'' \to 0$ it follows that if the second derivative f'_x is known then

$$\gamma^2 \rightarrow \left((\delta' + g')/f'_x\right)^2 \qquad (\text{mod } P_0) \tag{7.3.22}$$

and if the estimate h'' is known then

$$\gamma^2 \rightarrow \left((\delta' + g')/h''\right)^2 \qquad (\text{mod } P_0) \tag{7.3.23}$$

From (7.3.15), (7.3.16), (7.3.17), (7.3.22) and the Lebesgue theorem it follows that

$$\rho_{13}^2 = (2k_n)^2 E_0 \left\{ \left(\frac{\delta' + g'}{f''_{x_n}}\right)^2 \right\} \tag{7.3.24}$$

δ' is deterministic, the expectation of g' is zero and the variance is σ'^2. Then from (7.3.24)

$$\rho_{13}^2 \rightarrow (2k_n)^2 \left(f'^2_{x_n}\right)^{-1} (\delta'^2 + \sigma'^2) \tag{7.3.25}$$

In a similar way, from (7.2.23), it follows that

$$\rho_{13}^2 \rightarrow (2k_n)^2 (h''^2)^{-1} (\delta'^2 + \sigma'^2) \tag{7.3.26}$$

In the case of Lemma 7.3.1

$$\rho^2 = (2k_n)^2 \sup \left(f'^2_{x_n}\right)^{-1} (\delta'^2 + \sigma'^2) \tag{7.3.27}$$

and

$$\rho^2 = (2k_n)^2 \sup (h''^2)^{-1} (\delta'^2 + \sigma'^2) \tag{7.3.28}$$

where the supremum is defined with regard to all possible values of variables which are present in the expressions.

From (7.3.16), (7.3.27) or (7.3.28) it follows that the error of the local Bayesian method

$$\rho(x_n, x^*) \leq 2k_n \sup (\varepsilon f'_{x_n})^{-1} \sqrt{(\delta'^2 + \sigma'^2)} \tag{7.3.29}$$

and

$$\rho(x_n, x^*) \leq 2k_n \sup (\varepsilon h'')^{-1} \sqrt{(\delta'^2 + \sigma'^2)} \tag{7.3.30}$$

Inequalities (7.3.29) and (7.3.30) mean that the local Bayesian method will not necessarily converge even for the convex twice differentiable functions. In this sense the special local Bayesian method may appear to be less good than the much more general global Bayesian method which converges to a minimum of any continuous function; see Theorems 4.2.9 and 4.2.10.

Inequalities (7.3.29) and (7.3.30) show that in order to keep the random part σ' of the error small we have to use a large step size q_0 see (7.3.8). However, from (7.3.5) and (7.3.6) it follows that the deterministic part of the error, especially for non-symmetric functions, will then be increased. This means that in order to minimize the deterministic part of the error we should keep the step size q_0 as small as possible. This contradiction is apparenrtly the most important negative factor explaining the poor convergence inside the region defined by (7.3.29) and (7.3.30).

The r.h. sides of inequalities (7.3.29) and (7.3.30) define the radius of a sphere (with the minimum at its centre) which will be reached by the method (7.2.26) after n iterations. It is easy to see that in the Bayesian case, where $2k_n$ should be equal to 1, the radius will not converge to zero. This means that by the minimization of the Bayesian risk function (7.2.13) we shall reach only some sphere around the minimum. The behaviour of the Bayesian method inside the sphere remains unclear. To provide the convergence of the method to the exact minimum some correction should be provided. The convenient way to do it is as follows. Outside the region the parameter $2k_n = 1$. When the boundary of the sphere is reached the parameters k_n and q_0 are changed in accordance with the usual rules of stochastic approximation to ensure the convergence to the exact minimum with probability 1, namely

$$2k_n = \begin{cases} 1, & \text{if } n \leq n', \\ (n - n')^{-(1-\alpha-\nu)}, & \text{if } n > \max(n', n'_{max}) \end{cases} \qquad (7.3.31)$$

where $q_0 = \gamma_0 \, n^{-\nu}$, $\alpha \geq 0$, $\nu \geq 0$,

$$\alpha + \nu < 0.5, \quad \nu - \alpha > 0, \quad \gamma_0 > 0.$$

Here n' is the first n when the boundary of region (7.3.29) or (7.3.30) is reached. The maximal number of iterations is restricted by n'_{max} to meet the convergence conditions of stochastic approximation.

The condition (7.3.31) means that outside the region (7.3.29) or (7.3.30) we are using a strictly Bayesian method in the sense that the risk function (7.2.13) is minimal. Inside the region the step length should be reduced following the convergence conditions of stochastic approximation.

In a stochastic approximation it is well known, see Wasan (1965), that for continuously differentiable functions the asymptotically optimal α and v in the minimax sense are

$$\alpha = 0,$$

$$v = 1/4.$$

However, from (7.3.8) it follows that small q_0 means a large random component of the error σ' and a correspondingly low efficiency of search. So in the local Bayesian methods we shall use much smaller v. For example, $v = 0.01$, $\alpha = 0.09$ and $\gamma_0 = 0.2$.

Usually we do not know the distance $\rho(x_n, x^*)$. The absolute value of the gradient f_x or its estimate h' can be regarded as a reasonable estimate of $\rho(x_n, x^*)$, because for convex functions the value of $|f_x|$ is a monotonic function of the distance from the minimum. If the noise is large we can ignore the deterministic component δ'. In such a case the boundary of the regions (7.3.29) and (7.3.30) can be estimated from the inequalities

$$|h'| \leq 2k_n \, \sigma'/(\varepsilon f'_{x_n}) \tag{7.3.32}$$

and

$$|h'| \leq 2k_n \, \sigma'/(\varepsilon h'') \tag{7.3.33}$$

The parameter n' in (7.3.31) will be defined as the first n when

$$|h'| \leq \sigma'/(\varepsilon h'') \tag{7.3.34}$$

Here $1 - \varepsilon$ is a Lipschitz constant of transformation T_1 which depends on the function to be minimized.

If (7.3.34) does not occur before $n = n'_{max}$ then we shall assume that $n = n'_{max}$, because otherwise the convergence conditions will not be satisfied.

EXAMPLE 7.3.1. Suppose that

$$f_x = x^2, \ l' = 1, \ l'' = 1.$$

Then

$$\delta' = ((x + q_0)^2 - (x - q_0)^2)/(2q_0) - 2x = 0 \tag{7.3.35}$$

and

$$\delta'' = ((x + q)^2 - 2x^2 + (x - q)^2)/q^2 - 2 = 0 \tag{7.3.36}$$

From (7.3.8), (7.3.27) and (7.3.35)

$$\rho^2 = (2k_n)^2 \sigma^2/(4q_0^2 \, l') = k_n^2 \sigma^2/q_0^2. \tag{7.3.37}$$

If $k_n = k$ then the Lipschitz constant $1 - \varepsilon$ of transformation T_1 can be defined from (7.2.24) by the condition

$$\sup_{\substack{x_i \in R \\ x_j \in R}} \frac{|x_i - 2kx_i - (x_j - 2kx_j)|}{|x_i - x_j|} = 1 - \varepsilon \tag{7.3.38}$$

From here

$$\varepsilon = 2k \tag{7.3.39}$$

From (7.3.16), (7.3.37) and (7.3.39) the error of the Bayesian method

$$\rho(x_n, x^*) \le \sigma/(2q_0) \tag{7.3.39.1}$$

If the upper bound of the error does not exceed, for example 0.1, then from (7.3.39.1)

$$q_0 \ge 5/\sigma. \tag{7.3.40}$$

From (7.3.8) (7.3.34), (7.3.39) and $2k = 1$ it follows that n' is the first such n where

$$|h'| \le \sigma/(\sqrt{(2l')} \, q_0 h'') \tag{7.3.41}$$

In the progam LBAYES implementing the local Bayesian method, see section 9.19, the expression (7.3.41) is used to optimize not only quadratic but other functions as well, as an approximation of the inequality (7.3.16) which strictly speaking should define the switching moment n'. The reason is that usually the exact definition of the

Lifshitz constant $1 - \varepsilon$ and the distance $\rho(x_n, x^*)$, see expression (7.3.16), is a problem at least as difficult as the calculation of the optimal point x^*.

It is easy to see that the local Bayesian method defined by the expressions (7.2.24) to (7.2.30) can be considered as a version of the usual Newtonian method when the step length is reduced by the factor k_n. It is well known that the convergence of the Newtonian method depends very strongly on the second derivative. So we must take special care with its estimation. The estimate of the second derivative should be positive, otherwise condition (7.2.14) will correspond to the maximum of the risk function (7.2.12) rather than the minimum. Some stability of the estimate is also desirable in the sense that the deviation from the mean values should be restricted in some way. To satisfy these conditions we shall define the estimates α_n' as a weighted sum of the past estimates $\bar{\alpha}_i$, $i = 1, \dots , n$

$$\alpha_n' = \frac{L_0(1 - v)}{n^{1-v}} \sum_{i=1}^{n} \bar{\alpha}_i \, i^v \qquad (7.3.44)$$

where

$$\bar{\alpha}_i = \begin{cases} \alpha_i & \text{if } 10^{-6} \leq \alpha_i \leq 10^6 \\ 10^{-6} & \text{if } \alpha_i \leq 10^{-6} \\ 1 & \text{if } \alpha_1 \leq 10^{-2} \\ 10^6 & \text{if } \alpha_i \geq 10^6 \end{cases} \qquad (7.3.45)$$

The third condition in (7.3.45) provides a reasonably good initial estimate α_1'. We shall assign $L_0 = 1$. Here

$$\alpha_i = h'' q_0 = h_{x_i + q} - 2h_{x_i} + h_{x_i - q}$$

Condition (7.2.27) implies that the decision whether to go to the next point or not depends only on the results of the comparison of the observations at points x_n and $x_n - \beta_n$ respectively. The probability of a wrong decision can be reduced if the function is observed at each point not only once but, say, l and l_β times respectively. In the case where $l_n > 1$

$$l = l_n l_0$$

where l_0 is the number of repetitions in one iteration and l_n is the number of iterations when the observations are performed at the same point x_n.

Formula (7.3.44) was derived from the following considerations. The average of differences (7.2.19) can be expressed in the following way

$$\bar{h}" = 1/n \sum_{i=1}^{n} h"(i) \tag{7.3.46}$$

where $h"(i)$ is the observed value of $h"$ at the iteration i.
From (7.3.46), (7.2.19), (7.2.20) and (7.2.23)

$$\bar{h}" = 1/n \sum_{i=1}^{n} \frac{\bar{\alpha_i}}{q_0} \tag{7.3.47}$$

From (7.3.31) and (7.3.47) assuming $\gamma_0 = 1$

$$\bar{h}" = 1/n \sum_{i=1}^{n} \bar{\alpha_i} i^\nu . \tag{7.3.48}$$

Since the average of α_n may be expressed as

$$\alpha'_n = \bar{h}"/\bar{q}_0 \tag{7.3.49}$$

where $\bar{q}_0$ is the average of q_0.
From (7.3.49) and (7.3.47) the average value

$$\alpha'_n = 1/(n\bar{q}_0) \sum_{i=1}^{n} \bar{\alpha_i} i^\nu \tag{7.3.50}$$

We shall estimate the average q_0 as

$$\bar{q}_0 = (1-\nu)^{-1} n^{-\nu} \tag{7.3.51}$$

and then from (7.3.51) and (7.3.50) the average of $\bar{\alpha_i}$ which we shall denote as

$$\alpha'_n = \frac{1-\nu}{n^{1-\nu}} \sum_{i=1}^{n} \bar{\alpha_i} i^\nu . \tag{7.3.52}$$

Suppose that in the expression (7.3.31) the indices n denote only such iterations where $x_{n+1} \neq x_n$. Then the local Bayesian method defined by the expressions (7.2.29), (7.2.30), (7.3.31), (7.3.43) to (7.3.45) can be reduced to the usual method of a stochastic approximation and so will converge to the minimum under the same conditions.

7.4 Generalization of a multi-dimensional case

As usual different ways exist to generalize from the one-dimensional case to the case where $x = (x^1, \dots, x^m)$. A direct one is to regard the parameters d, a of the Bayesian statistical models (7.2.4) and (7.2.2) as the multi-dimensional vectors and matrices respectively. However such a model will contain too many unknown parameters such as the partial derivatives of the first and second order which must be estimated from a relatively small number of observations. Also, in this case we shall lose the simplicity which is the main advantage of the local Bayesian methods when comparing them with the global ones.

Another way to generalize is to use the same direction of search as in the usual methods of stochastic approximation and to apply the Bayesian approach only to optimize the step size. Here we use the Bayesian methods only to consider the problem of an optimal step size, which is less conveniently considered in the usual framework of the stochastic approximation. In this case we need to estimate only the partial derivatives of the first order to get the direction of search. The second derivative along the line of search can be estimated directly using some additional observations performed on this line.

When the direction $s = s_n$ is fixed then the multi-dimensional search is reduced to the one-dimensional search. The only variable is the distance from the initial point x_n.

Put

$$f_s(\beta) = f(x - \beta s) \tag{7.4.1}$$

and

$$x_{n+1} = x_n - \beta_n s_n \tag{7.4.2}$$

where

$$s_n = f'_{x_n} / \| f'_{x_n} \| \tag{7.4.3}$$

$$f'_{x_n} = \left(\frac{\partial f(x_n)}{\partial x^1} \ \cdots \ \frac{\partial f(x_n)}{\partial x^m} \right) \tag{7.4.4}$$

Then from (7.2.28) the optimal length of step

$$\beta_n = 2k_n f_{s_n}(0)/(f'_{s_n}(0)) \tag{7.4.5}$$

where

$$f_{s_n}(0) = df_{s_n}(0)/d\beta$$

and

$$f'_{s_n}(0) = d^2 f_{s_n}(0)/d\beta^2 > 0.$$

Here $df_s(0)/d\beta$ means the derivative of $f_s(\beta)$ at the point $\beta = 0$. The gradient along the fixed direction $s_n = (s_n^1, \ldots, s_n^m)$ can be expressed as a sum

$$f'_s(0) = \sum_{i=1}^{m} (\partial f(x_n)/\partial v_i) \ \partial v_i/\partial \beta \tag{7.4.6}$$

where

$$v_i = x_n^i - \beta s_n^i . \tag{7.4.7}$$

From here

$$\partial v_i/\partial \beta = -s_n^i \tag{7.4.8}$$

and

$$\partial f(x_n)/\partial v_i = \partial f(x_n)/\partial x^i . \tag{7.4.9}$$

From (7.4.6) to (7.4.9) the gradient along the direction s is

$$f'_s(0) = -\sum_{i=1}^{m} (\partial f(x_n)/\partial x^i)s^i \tag{7.4.10}$$

Suppose that the direction s is the direction of an anti-gradient

$$s^i = s_n^i = -\partial f(x_n)/\partial x^i / \|f_{x_n}\|, \ i = 1, \ldots, m . \tag{7.4.11}$$

From (7.4.10) and (7.4.11)

$$f'_{s_n}(0) = \|f_{x_n}\|. \tag{7.4.12}$$

From (7.4.5) and (7.4.12)

$$\beta_n = 2k_n \, \|f_{x_n}\| / f'_{s_n} \, . \tag{7.4.13}$$

Denote the i-th co-ordinate of the step length β_n along the direction s_n as β_n^i. Then from (7.4.13) and (7.4.5)

$$\beta_n^i = \beta_n s_n^i = 2k_n \, \frac{\partial f(x_n)/\partial x^i}{f''_{s_n}(0)} \tag{7.4.14}$$

Denote $h_x = f_x + g_x$, where g_x is a noise with zero expectation and standard deviation σ and $x = (x^1, \dots x^m)$.

Since the partial derivatives are unknown we shall use their estimates

$$h' = (h'_1, \dots , h'_m)$$

where

$$h'_i = (h^i_{q_0} - h^i_{-q_0})/2q_0 \tag{7.4.15}$$

$$h^i_{q_0} = h_{(x^1_n,\dots,x^i_n + q_0,\dots,x^m_n)} \tag{7.4.16}$$

and

$$h^i_{-q_0} = h_{(x^1_n,\dots,x^i_n - q_0,\dots,x^m_n)} \tag{7.4.17}$$

where

$$h_{x_n} = f(x_n) + g_{x_n} \, . \tag{7.4.18}$$

Here

$$x_n = (x^1_n, \dots , x^m_n).$$

The second derivative f''_s along the direction s_n will be estimated by the corresponding difference of the second order calculated using the results of the special observations made at the initial points x_n and also at two other points located on the line defined by the direction h' and placed symmetrically in relation to x_n.

In such a case the estimate of the second derivative along the direction h' can be expressed as

$$h" = \left(h_{x_n + qs_n} - 2h_{x_n} + h_{x_n - qs_n}\right)/q^2 \tag{7.4.19}$$

where

$$q = \sqrt{q_0} \tag{7.4.20}$$

and

$$s_n = h'/\|h'\|.$$

By the substitution of h' for f_{x_n} and $h"$ for $f_{x_n}"$ from (7.4.14)

$$\beta_n^i = 2k_n\, h_i'/h". \tag{7.4.21}$$

Suppose that $q^2 = q_0$. Then from (7.4.21) it follows that

$$\beta_n^i = k_n\, \Delta_n^i/\alpha_n \tag{7.4.22}$$

where

$$\Delta_n^i = h'_i\, 2q_0 \tag{7.4.23}$$

and

$$\alpha_n = h"q_0 \tag{7.4.24}$$

Parameter k_n can be calculated as in the one-dimensional case from the expressions

$$2/k_n = \begin{cases} 1 & \text{if } n \leq n' \\ (n-n')^{-(1-\alpha-\nu)} & \text{if } n > \max\,(n', n_{max}) \end{cases} \tag{7.4.25}$$

where $\alpha\nu$ are the same as in (7.3.31) and n' is the first n when the boundary of the region of slow convergence is crossed. This region in the multi-dimensional case will be defined in the next section, considering the convergence, see inequality (7.5.32).

Expressions (7.4.21) and (7.4.22) are the direct extensions to the multi-dimensional case of the corresponding one-dimensional expressions (7.2.29) and (7.2.30). So, the multi-dimensional version of the local Bayesian algorithm is the sequence

$$x_{n+1} = \begin{cases} x_n - \bar{\beta}_n & \text{if } h_{x_n - \bar{\beta}_n} \leq h_{x_n} + \varepsilon_n \\ x_n & \text{if } h_{x_n - \bar{\beta}_n} > h_{x_n} + \varepsilon_n \text{ and } l_n < L \end{cases} \qquad (7.4.26)$$

where

$$\bar{\beta}_n = (\beta_n^1, \dots, \beta_n^m). \qquad (7.4.27)$$

From (7.4.22)

$$\beta_n^i = k_n \Delta_n^i / \alpha_n. \qquad (7.4.28)$$

From (7.2.32)

$$\varepsilon_n = 0.9 \ \sigma / \sqrt{(\pi l)}. \qquad (7.4.29)$$

l_n shows how many times the lower inequality in condition (7.4.26) has occurred, l is the number of observations at the point x_n.

7.5   Convergence in the multi-dimensional case

Put

$$\rho^2(x_i, x_j) = E\{\|x_i - x_j\|^2\}. \qquad (7.5.1)$$

If x_i, x_j are fixed

$$\rho^2(x_i, x_j) = \|x_i - x_j\|^2. \qquad (7.5.2)$$

We shall regard

$$x_i = x_j$$

if

$$x_i(\omega z) = x_j(\omega) \pmod{P_0}. \qquad (7.5.3)$$

From (7.4.15)

$$h'_i = f_i + \delta'_i + g'_i \tag{7.5.4}$$

where

$$\delta'_i = f_i - f'_i(x_n) \tag{7.5.5}$$

and

$$f'_i = \frac{f(x_n^1, \ldots, x_n^i + q_0, \ldots, x_n^m) - f(x_n^1, \ldots, x_n^i - q_0, \ldots, x_n^m)}{2q_0} \tag{7.5.6}$$

$f'_i(x_n) = \partial f / \partial x^i$ at the point $x = x_n$,

$$g'_i = \frac{g_{(x_n^1, \ldots, x_n^i + q_0, \ldots, x_n^m)} - g_{(x_n^1, \ldots, x_n^i - q_0, \ldots, x_n^m)}}{2q_0} \tag{7.5.7}$$

Here g'_i is a random variable with zero expectation and standard deviation

$$\sigma'_i = \sigma / (\sqrt{(2l')} q_0). \tag{7.5.8}$$

The second derivative f''_{x_n} along the direction $h' = (h'_1, \ldots, h'_m)$, its estimate and, also, the deterministic and the random errors δ'' and g'', respectively, are defined by the same expressions (7.3.9) to (7.3.14) as in the one-dimensional case.

The distance between the transformations corresponding to (7.4.14) and (7.4.21) can be defined as

$$\rho^2 = \sum_{i=1}^{m} \rho_i^2 \tag{7.5.9}$$

where

$$\rho_i^2 = (2k_n)^2 E\{\gamma_i^2\} \tag{7.5.10}$$

and

$$\gamma_i^2 = \left(\frac{f'_i}{f''_{x_n}} - \frac{h'_i}{h''}\right)^2 \tag{7.5.11}$$

LEMMA 7.5.1. *Suppose that f is twice differentiable and that there exist the integrable functions s_i, s and the positive number β such that*

$$|g'_i| \le s_i, \quad i = 1, \ldots, m \qquad (\text{mod } P_0) \tag{7.5.12}$$

$$\left|f''_{x_n} + \delta'' + g''\right|^{-1} \le s \qquad (\text{mod } P_0) \tag{7.5.13}$$

$$\inf f'_{x_n} = \beta > 0. \tag{7.5.14}$$

Then from $l'' \to \infty$ and $\delta'' \to 0$ it follows that

$$\rho^2 = (2k_n)^2 \, 1/f''_{x_n} \, (m\sigma^2 + \sum_{i=1}^{m} \delta'^2_i) \tag{7.5.15}$$

assuming that the second derivative f'_{x_n} is known, and

$$\rho^2 = (2k_n)^2 \, 1/h'' \, (m\sigma^2 + \sum_{i=1}^{m} \delta'^2_i) \tag{7.5.16}$$

assuming that the estimate h'' of the second derivative is known.

Proof. From (7.5.11), (7.3.9) and (7.5.4)

$$\gamma^2 = \left(\frac{f'_i}{f''_{x_n}} - \frac{f'_i + \delta'_i + g'_i}{f''_{x_n} + \delta'' + g''}\right)^2 \tag{7.5.17}$$

From (7.5.17) and from $\delta'' \to 0$, $\sigma'' \to 0$

$$\gamma^2_i \to \left((\delta'_i + g'_i)/f''_{x_n}\right)^2 \qquad (\text{mod } P_0) \tag{7.5.18}$$

and replacing the second derivative by its estimate h''

$$\gamma^2_i \to \left((\delta'_i + g'_i)/h''\right)^2 \qquad (\text{mod } P_0) \tag{7.5.19}$$

From (7.5.10) to (7.5,13), (7.5.18) and from the Lebesgue theorem it follows that

$$\rho^2_i \to (2k_n)^2 E \left\{\left(\frac{\delta'_i + g'_i}{f''_{x_n}}\right)^2\right\} \tag{7.5.20}$$

Since δ'_i is deterministic, the expectation of g'_i is zero and the variance of g'_i is σ'^2, then from (7.5.20)

$$\rho_i^2 \;\to\; (2k_n)^2 \; 1/f'^2_{x_n} \; (\delta'_i + \sigma'^2). \tag{7.5.21}$$

From here and from (7.5.9)

$$\rho^2 \;=\; (2k_n)^2 \; 1/f''^2_{x_n} \; (m\sigma^2 + \sum_{i=1}^{m} \delta'^2_i) \tag{7.5.22}$$

In a similar way

$$\rho^2 \;=\; (2k_n)^2 \; 1/h'' \; (m\sigma^2 + \sum_{i=1}^{m} \delta'^2_i) \tag{7.5.23}$$

From (7.3.16), (7.5.22) and (7.5.23) the square of the error of a local Bayesian method is

$$\rho^2(x_n, x^*) \;\leq\; (2k_n)^2 \; \sup \; 1/(\varepsilon^2 f'^2_{x_n}) \; (m\sigma^2 + \sum_{i=1}^{m} \delta'^2_i) \tag{7.5.24}$$

and

$$\rho^2(x_n, x^*) \;\leq\; (2k_n)^2 \; \sup \; 1/(\varepsilon^2 h''^2) \; (m\sigma^2 + \sum_{i=1}^{m} \delta'^2_i) \tag{7.5.25}$$

By the substitution of $\|h'\|^2$ for $\rho^2(x_n, x^*)$ and neglecting the deterministic part of error δ'_i, $i = 1, \ldots , m$ from (7.5.25) we have

$$\|h'\|^2 \;\leq\; (2k_n)^2 \; \sup \; 1/\varepsilon^2 \; \frac{m\sigma'^2}{h''^2} \tag{7.5.26}$$

Since outside the region (7.5.26) parameter $2k_n$ is supposed to be equal to 1, then from (7.5.26)

$$\|h'\|^2 \;\leq\; \sup \; \frac{m\sigma'^2}{\varepsilon^2 h''^2} \tag{7.5.27}$$

EXAMPLE 7.5.1. Suppose that

$$f_x = \sum_{i=1}^{m} x_i^2, \quad l' = 1, \quad l'' = 1, \quad k_n = k.$$

Then

$$\delta_i' = 0 \quad \text{and} \quad \delta_i'' = 0.$$

In a way similar to that in example 7.3.1 we have

$$\varepsilon = 2k. \tag{7.5.28}$$

Then from (7.5.27) and (7.5.28) assuming $2k = 1$

$$\|h''\|^2 \leq \frac{m\sigma'^2}{h''^2} \tag{7.5.29}$$

and

$$\|h'\| \leq \sqrt{m}\, \sigma'/h''. \tag{7.5.30}$$

We shall use this definition of the region of slow convergence when the Lipschitz constant $1 - \varepsilon$ is not known.

From (7.5.8), (7.5.30), (7.4.24) and (7.3.44)

$$\|h'\| \leq \sqrt{(m/(2l'))}\, \sigma/\alpha_n'. \tag{7.5.31}$$

Here l' is the number of observations made at each of the points $(x_n^1, \ldots, x_n^i + q_0, \ldots, x_n^m)$ and $(x_n^1, \ldots, x_n^i - q_0, \ldots, x_n^m)$, $i = 1, \ldots, m$.

From (7.4.23) and (7.5.31)

$$\|\Delta_n\| \leq q_0 \sqrt{(2m/l')}\, \sigma/\alpha_n'. \tag{7.5.32}$$

Suppose that in the expression (7.4.25) the index n denotes only those iterations where $x_{n+1} \neq x_n$. Then the local Bayesian method defined by expressions (7.4.25) to (7.4.29) and (7.5.32) can be reduced to the usual method of a multi-dimensional stochastic approximation considered by Blum (1954).

Consequently the local Bayesian method will converge with probability 1 under the same conditions which are:

ASSUMPTION 7.5.33. The function f_x is continuous with the continuous derivatives of the first and second order.

ASSUMPTION 7.5.34. The derivatives of the second order are bounded.

ASSUMPTION 7.5.35. The variance σ is bounded.

ASSUMPTION 7.5.36. Suppose (without loss of generality) that $f_x = 0$ if $x = 0$ and $f_x > 0$ if $x \neq 0$. Then for any positive ε there exists the positive $\rho(\varepsilon)$ such that from $\|x\| \geq \varepsilon$ it follows that $f_x \geq \rho(\varepsilon)$ and $\|D(x)\| \geq \rho(\varepsilon)$ where $D(x)$ is the matrix of derivatives of the second order.

The multi-dimensional stochastic approximation method is usually defined as the sequence

$$x_{n+1} = x_n - (a_n/c_n)y_n \tag{7.5.37}$$

where

$$\lim_{n \to \infty} c_n = 0 \tag{7.5.38}$$

$$\sum_{n=1}^{\infty} a_n = \infty \tag{7.5.39}$$

$$\sum_{n=1}^{\infty} a_n c_n < \infty \tag{7.5.40}$$

$$\sum_{n=1}^{\infty} (a_n/c_n)^2 < \infty \tag{7.5.41}$$

Denote

$$y_n = \Delta_n = (\Delta_n^1, \ldots, \Delta_n^m) \tag{7.5.42}$$

$$c_n = q_0 \tag{7.5.43}$$

and

$$a_n/c_n = k_n/\alpha'_n. \tag{7.5.44}$$

Suppose that the index n denotes only those iterations where $x_{n+1} \neq x_n$. Consider the case when $n \geq n'_{max}$. Neither of these assumptions are important in the asymptotic sense, because they exclude only a finite number of iterations.

From (7.4.2), (7.4.22), (7.4.25) and (7.5.37)

$$a_n/c_n = (n - n')^{-(1-\alpha-\nu)} / \alpha'_n. \tag{7.5.45}$$

From (7.3.31) and (7.5.43)

$$c_n = \gamma_0 \, n^{-\nu}, \; \nu > 0, \; \alpha \geq 0, \; \alpha + \nu < 1/2, \; \nu - \alpha > 0, \; \gamma_0 > 0. \tag{7.5.46}$$

From (7.3.44) nd (7.3.45) there exist α' and α'' such that

$$0 < \alpha' \leq \alpha'_n \leq \alpha''. \tag{7.5.47}$$

Then conditions (7.5.38) to (7.5.41) hold, because from (7.5.46)

$$\lim_{n \to \infty} c_n = \lim_{n \to \infty} \gamma_0 \, n^{-\nu} = 0, \text{ if } \nu > 0, \; \gamma_0 > 0. \tag{7.5.48}$$

From (7.5.45) and (7.5.57)

$$\sum_{n=1}^{\infty} a_n = \sum_{n=n'+1}^{\infty} \frac{\gamma_0 n^{-\nu}}{(n - n')^{1-\alpha-\nu} \, \alpha'_n} + c$$

$$\geq \sum_{n=1}^{\infty} \gamma_0 / (n^{1-\alpha} \, \alpha') = \infty \text{ if } \alpha \geq 0. \tag{7.5.49}$$

From (7.5.45), (7.5.46) and (7.5.47)

$$\sum_{n=1}^{\infty} a_n c_n = c' + \sum_{n=n'+1}^{\infty} \frac{\gamma_0^2 n^{-2\nu}}{(n-n')^{1-\alpha-\nu}} = c' + \sum_{k=1}^{\infty} \frac{\gamma_0^2 (k+n')^{-2\nu}}{k^{1-\alpha-\nu}}$$

$$\leq c_0' + \gamma_0^2 \sum_{k=1}^{\infty} 1/k^{1-\alpha+\nu} < \infty, \text{ if } \alpha < \nu; \; k = n - n',$$

and

$$\sum_{n=1}^{\infty} (a_n/c_n)^2 = c + \sum_{n=1}^{\infty} \left(\frac{1}{(n-n')^{1-\nu-\alpha}}\right)^2 = c' + \sum_{k=1}^{\infty} \left(\frac{1}{k^{1-\nu-\alpha}}\right)^2 < \infty$$

if $\alpha + \nu < 1/2$. \hfill (7.5.50)

This means that under conditions (7.5.33) to (7.5.36) the local Bayesian method will converge to the minimum with probability 1.

The convergence of a method is, of course, a desirable property. However, if the number of iterations is small, the minimization of the maximal deviation ρ^2 as defined by (7.5.16) may be interesting.

Suppose that

$$\delta'_i = q_0, \ l' = 1. \tag{7.5.51}$$

From (7.5.8)

$$\sigma' = \sigma/\sqrt{2} \ q_0. \tag{7.5.52}$$

From here and (7.5.16)

$$\rho^2 = A \left(\sigma^2/(2q_0^2) + q_0^2\right) \tag{7.5.53}$$

where

$$A = (2k_n)^2 m/h''.$$

The minimum of ρ^2 is at the point

$$q_0 = (\sigma/\sqrt{2})^{1/2}. \tag{7.5.54}$$

7.6 The local Bayesian algorithm

It is convenient to write all the expressions which define the local Bayesian algorithm in one place.

From (7.4.26)

$$x_{n+1} = \begin{cases} x_n - \bar{\beta}_n & \text{if } h_{x_n - \bar{\beta}_n} \leq h_{x_n} + \varepsilon_n \\ x_n & \text{if } h_{x_n - \bar{\beta}_n} > h_{x_n} + \varepsilon_n \text{ and } l_n < L \end{cases} \tag{7.6.1}$$

where, from (7.4.27)

$$\bar{\beta}_n = (\beta_n^1, \dots, \beta_n^m).$$

From (7.4.28) substituting α'_n for α_n

$$\beta_n^i = k_n \left(\Delta_n^i / \alpha'_n \right).\tag{7.6.2}$$

From (7.4.29)

$$\varepsilon_n = 0.9 \, \sigma \varepsilon_0 / (\sqrt{\pi l}).\tag{7.6.3}$$

Usually $l = 2m$ and ε_0 is the correction parameter, usually $\varepsilon_0 = 1$.
From (7.4.25)

$$2/k_n = \begin{cases} 1 & \text{if } n \le n' \\ (n - n')^{-(1-\alpha-\nu)} & \text{if } n > \max (n', n_{\max}) \end{cases}\tag{7.6.4}$$

Here n' is the first n which satisfies the condition

$$\|\Delta_n\| \le \sigma q_0 \sqrt{(m \, 2/l')}/\alpha'_n.\tag{7.6.5}$$

It is obvious that the point x_n will not necessarily remain inside the region of slow convergence estimated by the inequality (7.6.5). It can move out at any iteration.

As a result the step size will be decreasing not only inside, but sometimes also outside, the region of slow convergence, which is defined by the following relation:

$$\|f_{x_n}\| \le \sqrt{(m/(2l'))} \, \sigma / (f'_{x_n} q_0).\tag{7.6.6}$$

This relation can be derived from (7.5.31) by the substitution of the derivatives f'_{x_n} and f'_{x_n} for their estimates h' and α'_n / q_0.

There are several ways of correcting the estimates (7.6.5) or (7.6.6) of the region. One of them is simply to multiply the r.h.s of (7.6.5) or (7.6.6) by the factor $k' < 1$. Then from (7.6.5)

$$\|\Delta_n\| \le k' q_0 \sqrt{(m \, 2/l')}) \cdot \sigma / \alpha'_n.$$

Another way is to allow the possibility of stopping the decrease of k_n if we are clearly outside the region of slow convergence

$$\|\Delta_n\| > k'' q_0 \sqrt{(m \, 2/l')} \, \sigma / \alpha'_n$$

where $k'' > k'$, for example $k'' = 0.5$ and $k' = 0.25$.

From (7.3.44)

$$\alpha'_n = \frac{1-v}{n^{1-v}} \sum_{i=1}^{n} \overline{\alpha}_i \, i^v \tag{7.6.7}$$

From (7.3.45)

$$\overline{\alpha}_i = \begin{cases} \alpha_i & \text{if } 10^{-6} \le \alpha_i \le 10^6 \\ 10^{-6} & \text{if } \alpha_i \le 10^{-6} \\ 1 & \text{if } \alpha_1 \le 10^{-2} \\ 10^6 & \text{if } \alpha_i \ge 10^6 \end{cases} \tag{7.6.8}$$

From (7.4.23) and (7.4.15)

$$\Delta_n^i = h_{(x_n^1,\dots,x_n^i + q_0,\dots,x_n^m)} - h_{(x_n^1,\dots,x_n^i - q_0,\dots,x_n^m)} \tag{7.6.9}$$

From (7.4.24), (7.4.19) and (7.4.20)

$$\alpha_n = h_{x_n + qs_n} - 2h_{x_n} + h_{x_n - qs_n}, \tag{7.6.10}$$

$$q = \sqrt{q_0} \tag{7.6.11}$$

and

$$s_n = \Delta_n / \|\Delta_n\|. \tag{7.6.12}$$

Here the result of the observation at the point x_n is denoted by h_{x_n}.
We can observe only the sum of the function f_{x_n} and the noise g_{x_n} so

$$h_{x_n} = f_{x_n} + g_{x_n} \tag{7.6.13}$$

where the expectation of g_{x_n} is zero and the standard deviation is σ.
If σ^2 is not known, its estimate σ_0^2 is used

$$\sigma_0^2 = \frac{A_0 K}{K-1} \left(1/K \sum_{i=1}^{K} (f_{x_n} + g_{x_n}^i)^2 - (1/K \sum_{i=1}^{K} (f_{x_n} + g_{x_n}^i)^2) \right). \tag{7.6.13.1}$$

Here $A_0 > 0$ is the correction parameter, $g_{x_n}^i$ is the i-th sample of noise and K is the number of samples. Since the variance of noise is assumed to be independent of x_n, the samples at different x_n are combined.

From (7.3.31)

$$q_0 = \gamma_0 n^{-\nu}, \ \gamma_0 > 0 \tag{7.6.14}$$

where

$$\nu > 0, \quad \nu + \alpha < 1/2, \ \nu - \alpha > 0, \ \alpha \geq 0. \tag{7.6.15}$$

In the case of linear constraints the projection operation P_A should be applied in the expression (7.6.1), see Ermoljev (1976)

$$x_{n+1} = \begin{cases} P_A (x_n - \bar{\beta}_n), & \text{if } h_{x_n - \bar{\beta}_n} \leq h_{x_n} + \varepsilon_n \\ x_n, & \text{if } h_{x_n - \bar{\beta}_n} > h_{x_n} + \varepsilon_n \end{cases} \tag{7.6.16}$$

7.7   Results of computer simulation

The following unimodal convex test function was considered by Zukauskaite (1987)

$$f(x) = \sum_{i=1}^{n} \left(\frac{x_i^3}{6} + \frac{x_i^2}{2} \right) \tag{7.7.1}$$

where $x_i \in [-1, 1]$.

Table 7.7.1 shows the results of computer simulation with different parameters.

Row No.			Column number						
			1	2	3	4	5	6	7
1	↑	m	2	2	5	5	10	10	20
2			0	0.05	0	0.05	0	0.05	0
3	Para-	α	0	0.05	0	0.05	0	0.05	0
4	met-	ε_0	1	1	1	1	1	1	1
5	ers	A_0	1	1	1	1	1	1	1
6		α'_n	B	B	B	B	B	B	B
7	↓	(Step reduction)	B	B	B	B	B	B	B
8	↑ Ave. error (in 10^{-3}	(Optimum point)	3.66	3.48	12.32	20.15	41.17	64.20	178.15
9	units) ↓	(Last point)	2.09	2.52	9.35	15.57	41.33	64.15	175.77

Row No.	8	9	10	11	12	13	14	15	16
				Column number					
1	20	2	2	2	2	2	2	2	2
2	0.05	0.25	0.05	0.25	0.05	0.05	0.05	0.05	0.25
3	0.05	0.25	0.05	0.05	0.05	0.05	0.05	0.05	0.25
4	1	10^6	10^6	10^6	1.0	10^6	10^6	2.0	10^6
5	1	10^6	10^6	10^6	1.0	10^6	1.0	2.0	10^6
6	B	1.0	1.0	B	B	B	B	B	0.4
7	B	S.A.	S.A.	S.A.	S.A.	S.A.	B	B	B
8	231.01	52.24	70.44	6.98	16.32	5.74	6.77	5.93	10.50
9	231.98	45.89	64.31	24.15	13.61	3.93	7.90	4.40	7.13

Table 7.7.1

Relationship of average error and parameters of the method

The best results were obtained by the local Bayesian algorithm defined in section 7.6 with parameters $v = \alpha = 0$, $\varepsilon_0 = A_0 = 1.0$ (see columns 1, 3, 5, 7). Column 9 corresponds to the classical algorithm of a stochastic approximation. Columns 10 to 16 show the results of gradual change from the Bayesian algorithm (Column 2) to a stochastic approximation procedure of usual type.

The noise corrupting $f(x)$ was Gaussian with zero expectation and the standard deviation was defined by the expression $\sigma = 0.046\sqrt{(m/2)}$ to keep the signal/noise ratio approximately constant.

The average deviation from the minimum was obtained from twenty random runs.

The last row corresponds to the points obtained in the last iterations. The second row from the bottom shows the average value of the function $f(x)$ at the point where the minimal function value is observed. The last iteration was usually a better one.

Judging from the results of simulation the best algorithm is the local Bayesian one with parameters corresponding to columns 1, 3, 5, 7 derived ignoring convergence conditions. The best algorithm with convergence conditions is shown in columns 2, 4, 6, 8.

The Bayesian algorithm (see column 2) was about twenty times better than the usual algorithms of stochastic approximation (see column 9). This can hardly be explained by random factors.

The condition $\varepsilon_0 = 10^6$, see (7.6.3), means that there will be no repetition of iterations at the same point. The condition $A_0 = 10^6$, see (7.6.13.1), means that the step length will be reduced after each iteration as it is in the usual methods of stochastic approximation. The letters S.A. in the 7-th row means the same, only without the change of the estimate of variance, which follows from (7.6.13.1). The letter B in the 6-th and 7-th rows means that the Bayesian espressions were used for the estimation of the second derivative and for the reduction of step size.

The number of iterations in all cases was twenty. The number of function evaluations was correspondingly $20(4m + 2)$, because $2m$ evaluations were used for the gradient estimation, $2m$ for the function evaluation and 2 for the estimation of the second order derivative.

The final form of the Bayesian algorithm as presented in section 7.6 was derived as a result of theoretical considerations and the extensive computer simulation using different test functions, see Mockus et al (1987).

CHAPTER 8

THE APPLICATION OF BAYESIAN METHODS

8.1 Introduction

It often happens that the easiest way to explain the possibilities and the area of application of some mathematical methods is to show how they work in some real life examples. The investigation of such examples is also useful as an extension of the set of special test problems which are usually more simple but not representative as real engineering problems. The method can perform reasonably well on some textbook problem and fail on practical tasks. It is not unexpected, because numerical methods are sometimes tailored to fit the well know test problems. To tailor the methods to fit the more complicated problems is more difficult especially when the values of function cannot be expressed explicitly and may only be calculated algorithmically with possibly some errors.

Thus the investigation of real life problems is useful for both the users and the developers of a method. To the users it helps to understand the possibilities and limitations of the methods, and to the developers it shows the strong and weak points of the methods. In this chapter some practical examples using the methods of this book will be described. In some cases the analytical expression of the functions and constraints will be given. In more complicated cases the behaviour of functions will be only briefly outlined, referring the reader to the original papers for more details.

Since the examples represent a very wide area of apparently disconnected fields of application, the order of examples is more or less arbitrary.

8.2 The optimization of an electricity meter

This is the first example of the practical application of the Bayesian approach to a real problem of engineering design. The optimization was done in the early seventies (see Mockus et al (1962)) but the meter is still in production without any significant change in its basic configuration. The task was to design a more accurate electricity meter using the open configuration of the magnetic circuit which was less sensitive to deviations of the technological conditions. It was supposed, at the time, that only the closed magnetic circuit could provide the high acuracy of the meter. Unfortunately the closed circuit design is too sensitive to deviations of the technological conditions. Such deviations were, and still are, common in the factory environments. As a result the closed circuit configurations which were sufficiently accurate when produced under laboratory conditions were not good at all when produced on the production line.

The objective function was the maximal value of error. The nonlinearity of the magnetic circuit made the objective function multimodal. Figure 8.2.1 shows how the cost function depends on two parameters, because at the first stage of optimization the cost was the objective. Later the cost was considered as a restriction and the maximal error was minimized.

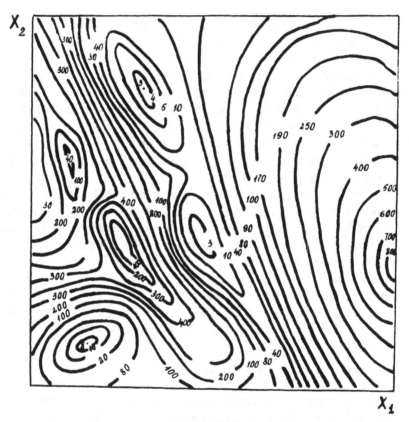

Figure 8.2.1

The cost function of an electricity meter as a function of two parameters

The multimodality problem was made easier using the parameters of the existing meter CO-444 as a starting point for the local search using a version of the gradient method. The Bayesian approach was used to test the hypothesis that the minimum found is a the global one. The idea of using the Bayesian approach to the optimization of the deterministic function was suggested by the unexpected empirical observation that the distribution of objective function values is surprisingly close to the lognormal distribution when the parameters of the electricity meter are uniformly distributed. It was shown by Mockus (1967) that the distribution will be close to lognormal if the objective function can be approximately represented as a product of functions which depend on a relatively small number of variables, when the total number of variables is large. In the case of the electricity meter, the number of variables was thirty.

So it was assumed that the *a priori* distribution of the objective function is lognormal and, under this assumption, the hypothesis that the global minimum is in the vicinity of the local minimum found using the parameters of the existing electricity meter CO-444 as the starting point, was tested.

Figure 8.2.2 shows the empirical (dotted line) and lognormal (continuous line) distributions when the parameters were estimated by the maximum likelihood method.

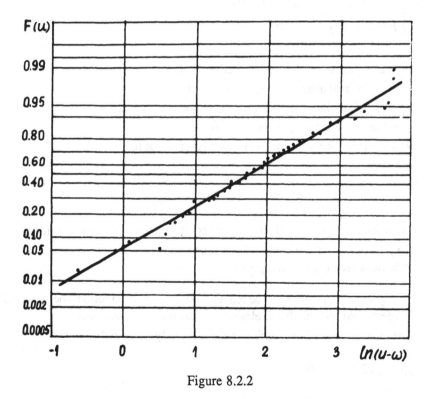

Figure 8.2.2

The distribution of the cost function of an electricity meter

The additional test of globality of the results was the repetition of the local minimization from different starting points. No better local minimum was found.

8.3   The optimization of vibromotors

The efficiency of vibromotors depends on the transfer of energy during the diagonal impact of the rigid bodies. To describe the process of diagonal impact the so called stereomechanical model can be used, see Ragulskiene (1974). We shall consider the special case of a mechanical model where the vibrating body 1 is moving the other body 2 in a straight line (see Didzgalvis et al.(1976)).

$$
\begin{cases}
m_1 y''_1 + v_3 f(y'_1 - y'_3 - y'_2) = 0 \\
(m_1 + m_3) v''_1 + v_3 = 0 \\
- v_3 f(y'_1 - y'_3 - y'_2) + u_3 + m_3 y''_3 = 0 \\
m_2 y''_2 + K y'_2 - u_3 = 0
\end{cases}
\qquad (8.3.1)
$$

Here

   m_1 and m_2 are the masses of the bodies 1 and 2,

   v_1 is the normal displacement of the bodies,

   y'_1 and y'_2 are the tangential velocities of bodies 1 and 2,

   y_3 is the tangential deformation of body 2,

   v_3 and u_3 are the normal and tangential reactions of the body 2,

where

   $v_3 = H_1 v'_1 - c_1 v_1$, $u_3 = H_3 y'_3 + c_3 y_3$,

   c_1, c_3 are the rigidities, H_1, H_3 are the coefficients of viscous friction,

   $f(\cdot)$ is the characteristic of dry friction and

   K is the coefficient of rolling friction.

The system of equations can be reduced to the normalized form if the surface of body 2 is described not only by its elasticity but also by its 'mass' m_3.

The characteristic of dry friction is approximately expressed as

   $f(v) = f_0\, 2/\pi\ \mathrm{arctg}\ (sv),$

where s is a sufficiently large number.

The initial conditions are defined by the momentary impulse on body 1 under the angle α

$$v'_1(0) = \cos \alpha, \quad y'_1(0) = \sin \alpha,$$

$$v_1(0) = y_1(0) = y_2(0) = y_3(0) = y'_2(0) = y'_3(0) = 0.$$

The end of the process T is the moment of separation of the masses. In the design of vibromotors the main variables are the angle of the impact α and the parameters of the materials. The angle α defines the initial conditions and the materials define the parameters $h_1, h_3, f_0, \gamma_1, \gamma_2$ of the system (8.3.1), where

$$\gamma_1 = c_3/c_1 \text{ and } \gamma_2 = m_1/m_2,$$

$$h_1 = H_1/(2\rho m_1), \quad h_3 = H_3/(2\rho m_3), \quad \rho = v(c_1/m_1).$$

The quality of a vibromotor can be expressed using the following functions:

a) the loss function in the contact in the normal direction

$$F_1 = 1/\rho \int_0^T v_3^* v'_1 \, d\tau, \quad v_3^* = 2h_1/\rho \, v'_1 + v_1, \quad \rho = \sqrt{(c_1/m_1)}$$

b) the loss function of friction in the contact

$$F_2 = f_0/\rho \int_0^T v_3^* (y'_1 - y'_3 - y'_2) \, d\tau$$

c) the function of useful work

$$F_3 = 2k/\rho^2 \int_0^T (y'_2)^2 \, d\tau$$

d) the efficiency

$$F_4 = \frac{(y'_2 \, (T))^2}{\gamma_2(1 - ((x'_1 \, (T))^2 - (y_1'(T))^2)/\rho^2)}$$

e) the velocity of body 2 at the end of the process

$$F_5 = y'_2(T)/\rho$$

f) the displacement of body 2 at the end of the process

$$F_6 = y_2(T)$$

g) the deformation of body 2 in the normal direction

$$F_7 = x_1(T).$$

The set of feasible points was the rectangular one:

$$0.2 \leq \gamma_1 \leq 10, \qquad\qquad 0.1 \leq h_1 \leq 0.8$$
$$0.1 \leq \gamma_2 \leq 0.5, \qquad\qquad 0.1 \leq h_3 \leq 0.8$$
$$0 \; < \alpha \; < \pi/2, \qquad\qquad 0.2 \leq f_0 \leq 1.$$

The functions F_1, F_7 can be expressed in the analytical form and depend on only two parameters h_1 and α where the ε-optimal angle is $\alpha = \pi/2 - \varepsilon, \varepsilon > 0$. So one-dimensional optimization can be used to define the optimal values of parameter h_1 which minimize the loss functions F_1 or F_7.

To define the other functions $F_2, \dots , F_6$ numerical integration of the system of differential equations is necessary. To perform it the well known method of prediction and correction was used.

Table 8.3.1 shows how the time of numerical integration of the system (8.3.1) depends on the error of integration R and the 'accuracy' of the mechanical model $l = m_1/m_3$, where m_3 is the 'mass' of surface. The accuracy of the model is highest when $m_3 \to 0$.

$R \setminus l$	10	100	500
10	1.5		
0.01	8.0		
0.05		25.0	114

Table 8.3.1

The relation of the C.P.U. time in seconds on the accuracy l of the mechanical model and the error R of numerical integration

Sectional views of functions $F_2, \dots , F_6$ are unimodal or at least 'bimodal', so it seems that local methods could be used to optimize functions $F_2, \dots , F_6$. However the presence of errors of numerical integration can transform even the unimodal function to the 'multimodal' one unless the accuracy of numerical integration and, consequently, the time of integration is very high.

The best results were achieved using the global methods and relatively low accuracy.

Parameters		Objective functions				
		F_2	F_3	F_4	F_5	F_6
h_1	global search	0.10	0.10	0.10	0.10	0.10
	local search	0.10	0.10	0.10	0.10	0.10
h_3	global search	0.10	0.10	0.10	0.10	0.10
	local search	0.38	0.10	0.10	0.10	0.10
γ_1	global search	0.88	1.52	0.93	0.93	1.53
	local search	0.41	1.05	0.73	0.74	1.12
γ_2	global search	0.50	0.50	0.50	0.50	0.50
	local search	0.50	0.50	0.50	0.50	0.50
α	global search	10^{-10}	0.92	0.82	0.99	0.91
	local search	10^{-10}	0.90	0.94	0.97	0.87
Obj.funct.	global search	10^{-12}	0.05	0.33	0.39	0.74
	local search	0.01	0.04	0.35	0.34	0.69
CPU time	global search	32.34	32.34	32.34	32.34	32.32
BESM-6 (min)	local search	150	78	92	100	86

Table 8.3.2

The results of the optimization

The co-ordinate optimization was done using the one-dimensional global method of Zilinskas (1975), see section 9.15. The results of global optimization are shown in Table 8.3.2. The local correction of the results was done by the variable metrics method (see Tieshis (1975)) in three stages: the first with $R = 0.01, l = 10$, the second with $R = 0.05, l = 100$ and the third with $R = 0.05, l = 500$.

The results of the last stage of local correction are also shown in Table 8.3.2.

8.4   The optimization of a shock-absorber

The mechanical object of two masses is considered. It is supposed that the shock is an instantaneous impulse and that the object can be represented by a system of linear differential equations. The shock-absorber contains damping and elasticity so it can be described as a linear function of deviation and velocity. The parameters of the shock-absorber should minimize the maximal deviation during the transitional process.

The mechanical model of the system, see Didzgalvis et al. (1978), is shown in Figure 8.4.1 where m_1 is the first mass, m_2 the second mass, H_1 and H_2 are two dampers, C_1 and C_2 are two elasticities.

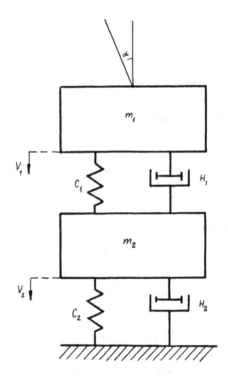

Figure 8.4.1

The mechanical model of a linear shock-absorber

The differential equations of the mechanical model:

$$m_1 v''_1 + H_1(v'_1 - v'_2) + c_1(v_1 - v_2) = 0$$

$$m_2 v''_2 + H_1(v'_2 - v_1) + c_1(v_2 - v_1) + p(v'_2, v_2) = 0 \qquad (8.4.1)$$

where v_1, v_2 are the co-ordinates of masses and $p(v'_2, v_2)$ is the linear action of the shock- absorber

$$p(v'_2, v_2) = H_2 v'_2 + c_2 v_2.$$

The shock defines the initial conditions

$$v_1(0) = v_2(0) = v'_2(0), \quad v'_1(0) = v = \cos \alpha, \; 0 < \alpha < \pi/2 \qquad (8.4.2)$$

The end of the transitional process is defined by the condition

$$H_1(v'_1 - v_2) + c_1(v_1 - v_2) = 0.$$

There are four parameters of the shock-absorber that can be optimized

$$x_1 = h_1 = H_1/(2m_2 \rho),$$

$$x_2 = h_2 = H_2/(2m_2 \rho),$$

$$x_3 = \gamma_1 = c_2/c_1,$$

$$x_4 = \gamma_2 = m_2/m_1, \qquad (8.4.3)$$

where

$$\rho = \surd(c_1/m_2).$$

The area of admissible values of x is defined as the intersection of the sets:

$$X = X_1 \cap X_2 \cap X_3 \qquad (8.4.4)$$

where

$$X_1 = \{x : 0.1 \le h_1 \le 0.8, \;\; 0.1 \le h_2 \le 0.8, \; 0.2 \le \gamma_1 \le 10, \; 2 \le \gamma_2 \le 10\} \qquad (8.4.5)$$

$$X_2 = \{x : \max_{0 \le t \le T} | 2h_2/\rho \; v'_2 + \gamma_1 v_2 | \le u_1\} \qquad (8.4.6)$$

$$X_3 = \{x : \max_{0 \leq \tau \leq T} [(2h_1/\rho \ (v'_1 - v'_2) + v_1 - v_2) \gamma_2] \leq u_2\} \tag{8.4.7}$$

where u_1 and u_2 are given constants, and $x = (x_1, x_2, x_3, x_4)$.

We should minimize the maximal deviation

$$F(x) = \max_{0 \leq \tau \leq T} |v_2(\tau)| . \tag{8.4.8}$$

To calculate the values of function $F(x)$ we must maximize the one-dimensional multimodal function $v_2(\tau)$. The convenient way to do it is by the one-dimensional global optimization method, see Zilinskas (1976).

Since the function $F(x)$ is not unimodal, and the observations are expensive, it is natural to use the global Bayesian method (see Mockus (1984[a])) after the reduction of the original problem with nonlinear constraints (8.4.4), (8.4.7) to the more convenient problem with rectangular constraints

$$\min_{x \in X_1} \{F(x) + b(s(q_1(x) - u_1) + s(q_2(x) - u_2))\} \tag{8.4.9}$$

where

$$s(q_i(x) - u_i) = \begin{cases} (q_i(x) - u_i)^2, & \text{if } q_i(x) > u_i \\ 0, & \text{if } q_i(x) \leq u_i \end{cases}$$

are the penalty functions, $i = 1, 2$.

$$q_1 = \max_{0 \leq t \leq T} \left| \frac{2h_2}{\rho} v'_2 - \gamma_1 v_2 \right|$$

$$q_2 = \max_{0 \leq t \leq T} \left(\frac{2h_2}{\rho} (v'_1 - v'_2) - v_1 - v_2 \right) \gamma_2 .$$

The results of the optimization are given in Table 8.4.1 for two sets of u_1 and u_2.

u_1	u_2	Optimal $F(x)$	Optimal x			
			x_1	x_2	x_3	x_4
0.15	8	0.0174	0.566	0.800	7.280	10.0
0.08	2	0.0312	0.140	0.681	0.452	10.0

Table 8.4.1

Results of optimization in the linear case

The case of a nonlinear shock-absorber can be considered in a similar way. The main difference in this case is that numerical instead of analytical integration of differential equations (8.4.1) should be used. The following nonlinear action of the shock-absorber was considered:

$$p(v'_2, v_2) = H_2 v'_2 + c_2 v_2 + c_3 v_2^3 \qquad (8.4.10)$$

The numerical integration of the corresponding differential equations was done using the standard FORTRAN routines, with step length 0.005, because with longer steps noticeable deviation from the analytical results was observed.

The results are shown in Table 8.4.2.

u_1	u_2	Optimal $F(x)$	Optimal x				
			x_1	x_2	x_3	x_4	x_5
0.02	0.7	0.14	0.20	0.10	0.20	2.09	180
0.02	2.4	0.017	0.163	0.146	0.44	10.0	10000
0.08	2.0	0.017	0.18	0.76	0.50	7.49	10000

Table 8.4.2

Results of optimization in the nonlinear case

In the nonlinear case the fifth variable

$$x_5 = \gamma_3 = c_3/c_1$$

was included because the additional parameter c_3 appeared.

8.5 The optimization of a magnetic beam deflection system

The function to be minimized $f(x)$ represents the aberration of the electron beam in a TV tube, see Grigas et al. (1980). The aberration depends on the ampere turns x_i of different sections of the beam deflection system, where $0 \le x_i \le 200$, $i = 1, \dots, 13$.

Aberration is the diameter of the light spot on the screen which defines the resolution of the tube. The aberration was estimated as the maximum of the deviations of the traces of four electron beams from the trace of the fifth central beam. The paths of electrons of all five beams were calculated using the following differential equations.

$$dx/dz = u,$$

$$dy/dz = v,$$

$$du/dz = \mu_0 k_0 \sqrt{(1 + u^2 + v^2)} \, (- uvH_x + (1 + u^2)H_y - vH_z)$$

$$dv/dz = \mu_0 k_0 \sqrt{(1 + u^2 + v^2)} \, (uvH_y - (1 + v^2)H_x + uH_z) \qquad (8.5.1)$$

Here $k_0 = \sqrt{(e/2m_e u)}$, and e, m_e are the charge and mass of the electron, u is the anode voltage, μ_0 is the magnetic permeability and H_x, H_y, H_z are the components of magnetic field strength (magnetic intensity). An efficient method of integration of the differential equations based on cubic splines was developed.

Since the coil consists of many sections the calculation of the magnetic strength H by the direct summing of the magnetic strengths of all sections is too complicated. So the coil was represented as a system of distributed three-dimensional currents. The optimization was done using the five algorithms of global optimization. In global optimization it is reasonable to use different algorithms based on different ideas and developed by different authors. In such a way we can obtain a deeper understanding of the problem and also compare the efficiency of the algorithms under consideration. Table 8.5.1 shows the corresponding results.

Algorithm No.	1	2	3	4	5
f_g	0.51	0.41	0.44	0.46	0.21
f_l	0.28	0.29	0.31	0.26	
N	403	399	380	308	233

Table 8.5.1

The results of minimization of the aberration of a TV tube using five different algorithms of global optimization

Algorithm 1 is the uniform deterministic search of LP type, by Sobol (1969), see section 9.11.

Algorithm 2 is the clustering algorithm by Törn (1978) with the number of initial points 10, see section 9.12.

Algorithm 3 is the adaptive Bayesian, by Mockus (1984) with the number of initial points 25, see section 9.9.

Algorithm 4 is the multi-dimensional algorithm by Zilinskas (1981) with parameters $K = 30$ and $K_L = 5$, see section 9.10.

Algorithm 5 is the co-ordinate Bayesian by Zilinskas with parameters $K = 6$, $N = 20$, $E_1 = E_2 = 0.001$, $i = 1, ... , 13$ and initial point $x = 35$, see section 9.15.

f_g is the result of global search,

f_l is the result of local search from the best point of global search,

N is the total number of observations, including global and local search.

The local search was done using the simplex method of Nelder and Mead, see Himmelblau (1972), with the initial length of the edge of the simplex equal to 40, see section 9.18.

The number of observations was divided between the global and the local search by the condition of equal sharing of CPU time. The exception was the co-ordinate optimization Algorithm 5, where no additional local optimization was needed because Algorithm 5 also provides the local minimum. The equal sharing of the CPU time means that the more complicated global algorithms such as adaptive Bayesian 3 were using a much smaller number of observations than the simpler algorithms such as the uniform search 1 or clustering 2. It explains why Algorithms 3 and 4 which are usually more efficient in the sense of numbers of observations did not behave much better that the simpler Algorithms 1 and 2.

The unexpected result was the success of the co-ordinate optimization Algorithm 5. To explain it the analysis of the structural characteristics of the function was carried out by the Shaltenis (1980) method, see section 9.20. The results are shown in Table 8.5.2.

Variable No.	13	12	11	1	2	3	4	5
Structural characteristics	20.8	15.8	9.1	6.9	6.7	5.1	4.5	2.9

Table 8.5.2

Structural characteristics of the variables

The analysis was based on the results of 150 observations by the LP grid by Sobolj (1969), see section 9.11.

The results of Table 8.5.2 show that following Shaltenis and Radvilavichute (1980) the 13-dimensional problem of optimization can be approximately reduced to eight problems of one-dimensional optimization, because the combined 'influence' (variance) of the remaining five variables and all possible combinations of variables does not exceed 30%. In such a case it is natural to expect that the co-ordinate optimization of the eight most 'influential' variables will be the most efficient algorithm.

8.6 The optimization of small aperture coupling between a rectangular waveguide and a microstrip line

It is well known, see Mashkovtsev et al (1966), that if the diameter of apertures coupling two lines is small in comparison with the wavelength then the coupling coefficient of the j-th aperture can be expressed as follows

$$c_j^\pm = \pm i\omega/2 \, \{(y_1 y_2)^{1/2} \, \mu_2 \, H_2^\mp(x_j) \, M_j \, H_1^+(x_j) \pm \varepsilon_2[E_2^\mp(x_j) \, n_j]P_j \, [E_1^+(x_j) \, n_j]$$

$$(8.6.1)$$

where the sign '−' corresponds to the return direction and the sign '+' to the straight direction, ω is the radial frequency, x_j is the co-ordinate of the j-th aperture, n_j is the normal to the j-th aperture and E_1^+, H_1^+ and $E_2^\mp, H_2^\mp$ are the normalized wave fields in the first and second lines respectively. μ_2, ε_2 are environmental parameters of the second line, y_1, y_2 are the specific admittances of the first and second lines, P_j is the coefficient of electrical polarization and M_j is the tensor of magnetic polarization of the aperture j.

The coupling system of N apertures can be characterized by the direct coupling factor S^+, the return coupling factor S^- and directivity D_f which are defined by the formulae

$$S_f^\pm = -20 \lg \left| \sum_{j=1}^{N} c_j^\pm \right| \qquad (8.6.2)$$

$$D_f = S_f^- - S_f^+. \qquad (8.6.3)$$

It was assumed that the number of apertures N was fixed, and that we should find such co-ordinates of apertures x_j and such coefficients P_j, M_j, of polarization which minimize the deviation from the desirable values of the characteristics S^+ and S^- and D.

We shall consider the case of four rectangular apertures, see Nikolaev et al (1984), and will use the well known (see Rao et al (1981)) expressions to define how the values of H, E, P and M depend on the geometrical parameters of the coupling such as the angle γ of waveguide slot, the length of the j-th slot L_j, the width of the i-th slot d_j and the distance between the centres of the j-th slot and the first slot z_j.

The objective function was defined as the maximal distance of the coupling factor S_f^+ from the fixed value S_0^-:

$$f(x) = \max_{f \leq f \leq f''} |S_f^+ - S_0^+| \tag{8.6.5}$$

The feasible set was defined by the condition that the minimal directivity should not be less than the prescribed value D_0:

$$D = \min_{f \leq f \leq f''} D_f \geq D_0. \tag{8.6.6}$$

Here f is the frequency and f' and f'' are the minimal and maximal frequencies, respectively.

The angle γ and the widths of the waveguide slots $d_j = d$ were fixed. To keep the apertures apart and to eliminate the dependence on the numbering of the apertures the following conditions were included

$$z_1 = 0,$$

$$z_{j+1} \geq z_j + d/\cos \gamma, \; j = 1, \dots, N-1,$$

$$z_N \leq z_{max}. \tag{8.6.7}$$

The lengths of the wave slots were restricted by the condition

$$L' \leq L_j \leq L'' \tag{8.6.8}$$

where L', L'' are minimal and maximal lengths.

We should minimize the multimodal function (8.6.5) with one convex constraint (8.6.6) and three linear constraints (8.6.7) in the rectangular area defined by (8.6.8). The variables $x = (x_1, \dots, x_7)$ where $x_i = L_i, i = 1, \dots, 4$, $x_{i+4} = z_{i+1} - x_i, i = 1, 2, 3, 4$.

To reduce the problem to minization in the rectangular area a penalty function of a special kind was introduced in the following way: The minimization with constraints

$$\min_{x \in A} f(x),$$
$$A = \{x : y(x) \geq 0\} \subset R^m \tag{8.6.9}$$

was reduced to the unconstrained minimization $\min_x f_0(x)$ where

$$f_0(x) = \begin{cases} f(x), & \text{if } x \in A. \\ f(x_A) + h(x, x_A), & \text{if } x \bar{\in} A. \end{cases} \qquad (8.6.10)$$

Here x_A is a mapping of x on A and $h(x, x_A)$ is an increasing function of the distance between x and x_A.

The function h is better than the usual penalty function in the sense that under some conditions function $f_0(x)$ will have no local minima outside the area A. This is a desirable property if we wish to find the global minimum by using some local optimization methods many times from different, not necessarily feasible, starting points.

In the case of linear constraints it is convenient to define the mapping x on A as the intersection of the border line of the inequality and the line connecting the point x and some feasible point. In this case for the constraint $\sum_{i=1}^{m} c_i x_i = b$ we have the following mapping

$$x_{iA} = (x_i - a_i)(b - \sum_{i=1}^{m} c_i a_i) / \sum_{i=1}^{m} a_i(x_i - a_i). \qquad (8.6.11)$$

Here

$$a_i = d/\cos \gamma, \ i = 1, \dots, m.$$

This penalty function was used for the last constraint from (8.6.7)

$$z_N = \sum_{i=N-1}^{2N-1} x_i \le z_{max}.$$

The optimization was performed in the following way:

Step 1: The maximization of the constraint function D under the conditions (8.6.7) and (8.6.8) by the method of global optimization UNT, see section 9.10. The number of initial points was 100, the total number of observations was 500, which is the maximum for UNT. The method located 16 local maxima not one of which was considered as satisfactory because the value of the function D was less then D_0. So local optimization was performed from the best points using the variable metrics procedure MIVAR4, see section 9.16. In this way several different feasible starting points were found and the local optimization of the objective function (8.6.5) was carried out using all of them as starting points. In this case the nonlinear constraint (8.6.6) was eliminated using the usual quadratic penalty function:

$r[\max (0, \varepsilon + D_0 - D)]^2$ where r is large and $\varepsilon > 0$ is small.

The maximization of $|S_f^+ - S_0^+|$ and the minimization of D_f where done simply by comparing the values of $|S_f^+ - S_0^+|$ and D_f for some fixed number of frequencies f. This number was smaller for the global optimization and larger for the local one because the procedures of variable metrics type are rather sensitive to even small errors of calculation. Figures 8.6.1 and 8.6.2 show how functions S_f^+ and D_f depend on the frequency f.

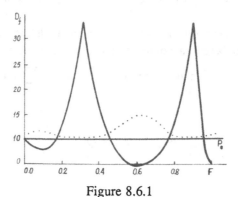

Figure 8.6.1

Relation of directivity D_f to normalized frequency $F = (f - f')/(f'' - f')$

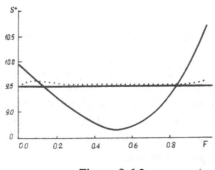

Figure 8.6.2

Relation of direct coupling factors S_f^+ to normalized frequency F

The continuous line corresponds to the parameters x fixed by experienced designers as an initial guess. The dotted line corresponds to the optimized parameters x. The horizontal lines correspond to the acceptable levels of S_f^+ and D_f.

8.7 The maximization of LSI yield by optimization of parameters of differential amplifier functional blocks

The deviation from the ratings of parameters which are set by the designer is the source of rejection of a substantial proportion of LSI because they do not meet the standards. One way of increasing the yield of LSI is to make the deviation as small as possible by using better and more expensive technologies. The other way is to optimize the ratings to reduce the percentage of rejects. Change of ratings during production is highly undesirable, so the convenient approach is Monte Carlo simulation when it is assumed that the deviations of parameters from their nominal values correspond to the same probability distributions, for example, the multivariate Gaussian one. In this case the objective function is the yield of LSI and it can be calculated only with some 'noise'. This noise can be decreased by numerous repetitions of the simulation process which is usually rather expensive.

Here, we shall consider the simple case of the optimization of only two parameters of the DA (differential amplifier) shown in Figure 8.7.1 (see Bashkis et al (1982)).

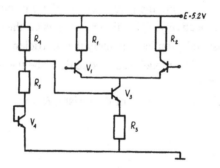

Figure 8.7.1

The differential amplifier; $R_1, \dots , R_5$ are resistors and $V_1, \dots , V_4$ transistors

It was supposed that the resistors could be subdivided into two parts, one with width x_1 and the other with width x_2. The fitness of the DA block was defined as a function of two parameters: the bias of zero U_0 and the lower level of the output voltage U^0. The block DA is considered fit if the following conditions hold

$$-2.10^{-3} \leq U_0 \leq 2.10^{-3}$$

and

$$4.23 \leq U^0 \leq 4.37 \qquad (8.7.1)$$

where the units of U_0 and U^0 are volts. Both parameters U_0 and U^0 are defined using the simplified model of transistors, the so called transconde model which, according to Bashkis et al (1984) provides sufficient accuracy. (The maximal deviation from the actual transfer characteristics was 3-4% at 60 times less CPU time in comparison with the well known transistor model of Ebers and Moll.)

$$U_0 = 1/2\, \alpha_3 \frac{\gamma_1}{R_3 + \gamma_3} \left[\frac{R_5 + \gamma_4}{R_4 + R_5 + \gamma_4} (E - \psi_4) - \psi_3 + \psi_4 \right)$$

$$\times \left(\frac{R_1}{R_2} - \frac{\gamma_1}{\gamma_2} + \frac{\alpha_1}{\alpha_2} - 1 \right) + \psi_1 + \psi_2 \right] \qquad (8.7.2)$$

$$U^0 = E - \frac{R_1}{R_3 + \gamma_3} \alpha_1 \alpha_3 \left[\frac{R_5 + \gamma_4}{R_4 + R_5 + \gamma_4} (E - \psi_4) - \psi_3 + \psi_4 \right] \qquad (8.7.3)$$

Here γ_i, ψ_i, $i = 1, \dots, 4$ are the parameters of the transconde representation of transistors $V_1, \dots V_4$ respectively. The dimension γ_i is Ω, the dimension of ψ is V and α_i is a non-dimensional parameter, namely the coefficient of current transfer.

The Gaussian statistical model was used to describe the parameters of resistors R_j and the transistors V_j.

The mean value of the resistance of the resistor R_j was defined as

$$\mu(R_j) = c_j R, \qquad (8.7.4)$$

and standard deviation as

$$\sigma(R_j) = 0.33\, h/d_j, \qquad (8.7.5)$$

where c_j is the number of squares laid in a row to make the resistor, R is the sheet resistivity (ohms per square), d_j and h are the nominal values and the standard deviation of the resistors width.

The mean values and the standard deviations of the transistor parameters are shown in Table 8.7.1.

The correlation coefficients of the parameters of resistors are shown in Table 8.7.2.

The violation of conditions (8.7.1) is not the only source of rejects. Another is defect of the silicon crystal. Rejects due to the silicon defects will be called 'catastrophic' rejects.

	Parameter					
Transistor	Mean			Standard deviation		
ψ	γ	α	$\sigma(\psi)$	$\sigma(\gamma)$	$\sigma(\alpha)$
V_1, V_2	0.74	23	0.992	8.10^{-3}	0.1	13.10^{-4}
V_3	0.76	15	0.992	8.10^{-3}	0.1	13.10^{-4}
V_4	0.72	30	–	8.10^{-3}	0.1	–

Table 8.7.1

Means and standard deviations of the transistor parameters

	R_i	γ_i	ψ_i	α_i
R_j	0.95	0.95	0	0
γ_j	0.95	0.95	0	0
ψ_j	0	0	0.8	0
α_j	0	0	0	0.7

Table 8.7.2

The correlation coefficients of the resistor and transistor parameters

The following formula, Valiev (1969), was used to estimate the influence of catastrophic defects on the yield

$$K = \exp(-0.4gS) \tag{8.7.6}$$

Here g is the density of the defects, S is the area of the silicon crystal, K is the proportion of fit circuits.

The area of the silicon crystal of a DA integrated circuit was calculated by the following formula

$$S = S_A + \sum_{j=1}^{5} \left[2\eta_j r_j (c_j d_j + 2r_j + d_j) + c_j d_j^2 \right] \tag{8.7.7}$$

where S is the area of the resistors together with the area between resistors and surrounding elements of the integrated circuit, S_A is the auxiliary area, η_j is the layout coefficient, j is the number of the resistor and r_j is the distance between the resistor j and other elements.

The objective function is the expected number N of good IC from a silicon wafer of area S_0:

$$f(x) = E(N) = S_0/S \ \exp(-0.4gS) \ M(S) \tag{8.7.8}$$

Here $M(S)$ is the proportion of IC which satisfies the conditions (8.7.1).

The widths of the resistors R_1, R_2, R_3 were denoted by x_1 and the widths of the resistors R_4 and R_5 were denoted by x_2, where

$$0 \le x_i \le 200\mu.$$

The optimization was carried out for two levels of technology. The lower level was described by the density of defects $g = 1.0$ mm^{-2} and the standard deviation of resistor widths $h = 1.5\mu$. The higher level was characterized by $g = 0.3$mm^{-2} and $h = 0.6\mu$. The other parameters were as follows: $S_A = 0.2$mm^2, $\eta = 0.8$, $r_j = 20\mu$. $c_1, c_2 = 2.3$, $c_3 = 1.3$, $c_4 = 32.7$ and $c_5 = 4.7$. The sheet resistance was $R = 200$ $\Omega/\square$ and the area of wafer was $S_0 = 4.5.10^3$mm^2.

The Monte Carlo simulation of the differential amplifier was done using a multi-dimensional Gaussian random number generator with the mean values, standard deviations and correlation coefficients of transistor and resistor parameters shown in Tables 8.7.1 and 8.7.2.

The results of the Monte Carlo simulation are shown in Figure 8.7.2 for the special case when $x_1 = x_2 = x$. The simulation was repeated 100 times and the average values plotted. The upper line corresponds to the higher technology and the lower line describes the lower technology. It is easy to see that the errors of the Monte Carlo simulation make the lower line 'multimodal'. It was shown by Bashkis (1984) that unimodality of the lower line can be provided by increasing the number of repetitions from 100 to 1000 which means 10 times more CPU time. This can be too much for the statistical optimization if the number of variables is large. It seems that a more efficient way is not to increase the repetitions but to regard the objective function as the multimodal one and to use the methods of global optimization In this case the adaptive Bayesian method BAYES1, see section 9.9, was used. This provided the optimal widths for the lower technology $x = (22, 45)$ and for the higher technology $x = (10, 33)$. The corresponding widths of resistors are given in microns.

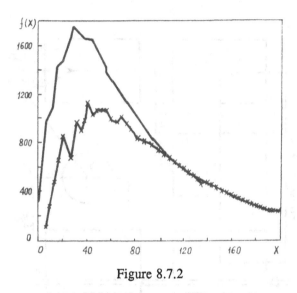

Figure 8.7.2

LSI yield as a function of resistor width

It is not unusual that in planning the production of IC for the lower level of technology the same parameters which were optimal for the higher technological level are fixed. This is a very simple but very inefficient way.

For example if, following such an approach, for the lower technology the widths $x = (10, 33)$ were chosen then the estimated yield would be 5500 acceptable IC. The optimal widths for the lower technology are $x = (22, 45)$ and the corresponding yield is 8400.

8.8 Optimization of technology to avoid waste in the wet-etching of printed circuit boards in iron-copper-chloride solutions

The processing plant for the electrochemical regeneration of iron-copper-chloride solution is considered. The idea is to get a dense sedimentation of copper on the cathode. However the recycling procedure cannot eliminate all waste: for example, inevitable entrainment of solution during the etching, the accumulation of excess copper chloride and the evaporation of water and hydrochloric acid. So it is important to set such parameters of the plant which minimize the waste and make the technology as clean as possible. The plant which we shall consider is shown in Figure 8.8.1 where ES is the etching set, IR is the intermediate reservoir, RS is the regeneration set, CS is the correction set for the etching solution, BS is the boiler and WS is the washing set to wash printed circuits (PC).

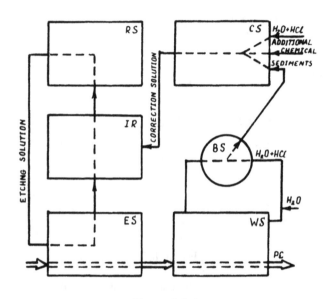

Figure 8.8.1

Block diagram of plant for printed circuit board production

It was proposed by Praparov et al (1967) to consider the problem of reducing the waste as the minimization of the cost function.

The cost function is defined as a result of the solution of equations of material balance of the plant for $20000m^2$ PC per year and includes the cost of materials and electrical power.

To keep the cathode sedimentation of copper dense during the process of electrochemical regeneration we should satisfy the following seven inequalities:

1) $0.016 \leq x_1 \leq 0.035$, where x_1 is iron chloride, $kg\ l^{-1}$.

2) $0.08 \leq x_2 \leq 0.165$, where x_2 is copper chloride, $kg\ l^{-1}$.

3) $0.003 \leq x_3 \leq 0.0069$, where x_3 is the copper etched from PC, $kg\ l^{-1}$.

4) $0.04 \leq x_4 \leq 0.07$, where x_4 is ammonium chloride, $kg\ l^{-1}$.

5) $0.03 \leq x_5 \leq 0.06$, where x_5 is hydrochloric acid, $kg\ l^{-1}$.

6) $1.0 \leq x_6 \leq 2.0$, where x_6 is the velocity of the flow of electrolyte in the space between the electrodes, cm sec^{-1}.

7) $0.005 \leq x_1 - 2x_3\ M_1/M_2 \leq 0.025$, where M_1 and M_2 are the molecular weight of iron chloride and copper respectively.

The last condition was considered using the penalty function. The minimal cost of m^2 of PC is 0.93 roubles. The optimal parameters are shown in Table 8.8.1.

x_1	x_2	x_3	x_4		
0.032	0.080	0.003	0.070	0.049	1.40

Table 8.8.1

Optimal parameters of PC technology

The global optimization was carried out by the uniform deterministic search of LP type by Sobolj (1968) and Dzemyda (1983), see section 9.11, using the analysis of structure by Shaltenis and Dzemyda (1982) and the grouping of variables by Dzemyda (1982).

The local optimization was done using the FLEXI method by Himmelblau (1972), see section 9.18. Fifty observations were made for the analysis of structure and 500 observations for LP optimization (Dzemyda et al (1984)).

8.9   The optimization of pigment compounds

The problem is to minimize the deviation of the colour of the pigment compound from some standard colour. The deviation of the compound from the standard can be defined in many different ways. We shall consider the two simplest. The first objective is the mean square difference of spectral reflectance coefficients corresponding to thirty-six wavelengths. The second is the difference in colours for a fixed light source.

It is well known that the difference of spectra can be calculated using the following formulae

$$f_1(x) = 1/L \sum_{k=1}^{\bar{L}} \left(r_k(x) - Q_k \right)^2, \quad L = 36$$

where

$$r_k(x) = a_k(x) - \sqrt{(a_k^2(x) - 1)}$$

$$a_k(x) = \frac{\sum_{i=1}^{n} x_i S_{ik} A_{ik}}{\sum_{i=1}^{n} x_i S_{ik}}, \quad A_{ik} = 0.5(R_{ik} + 1/R_{ik})$$

Here x_i is the density of the pigment i, $0 \le x_i \le 1$, $\sum_{i=1}^{m} x_i = 1$, $m = 9$, R_{ik} is the reflectance coefficient of the spectrum for pigment i at wavelength L_k. Coefficient R_{ik} is usually defined in the corresponding tables, see Barauskas et al. (1980).

S_{ik} is the diffusion of pigment at wavelength L_k,

Q_k is the reflectance coefficient of the standard of wavelength L_k, defined by the tables, see Barauskas et al. (1980).

The difference of colours for the fixed source of light with the co-ordinates $z = (z_1, z_2, z_3)$ where $z_1 = 96.8696$, $z_2 = 99.9994$, $z_3 = 112.1363$ was defined by the following formulae

$$f_2(x) = \left(\sum_{j=1}^{3} (w_j(r(x)) - w_j(Q))^2 \right)^{1/2}$$

where

$$r(x) = (r_1(x), \dots, r_{36}(x))$$

$$Q \; (Q_1, \dots, Q_{36})$$

$$w_1(r) = 25(100 \, Y_2(r)/z_2)^{1/3} - 16$$

$$w_2(r) = 13w_1(r) \, (uY(r)) - u(z))$$

$$w_3(r) = 13w_1(r) \, (uY(r)) - v(z))$$

$$r = (r_1, \dots, r_{36}), \quad z = (z_1, z_2, z_3)$$

$$Y(r) = (Y_1(x), \dots, Y_3(r))$$

$$Y_j(r) = \sum_{k=1}^{36} \psi_{jk} \, r_k,$$

$$u(Y) = \frac{4Y_1}{Y_1 + 15Y_2 + 3Y_3}, \quad v(Y) = \frac{9Y_2}{Y_1 + 15Y_2 + 3Y_3}$$

Nine pigments were considered:

1. Zinc oxide
2. Red iron oxide
3. Yellow iron oxide
4. Chromium oxide
5. Smoke-black
6. Orange lead glass
7. Dark green cobalt
8. Blue manganese
9. Ultramarine

The different standards for the colour were investigated.

The condition $\sum_{i=1}^{m} x_i = 1$ was eliminated using the multiplicative penalty function

$$F(x) = f(\tilde{x}) \exp \|\tilde{x} - x\|$$

where $\tilde{x}_i = \dfrac{x_i}{\sum_{j=1}^{m} x_j}$ and $\|\tilde{x} - x\| = \left(\dfrac{1}{\sum_{i=1}^{m} x_i} - 1\right)^2 \sum_{i=1}^{m} x_i^2$.

The results of optimization by four different methods are shown in Table 8.9.1.

In Table 8.9.1 the following algorithms were considered.

1. The uniform deterministic of Sobolj (1969), see section 9.11, with 3000 observations.

2. The clustering algorithm of Törn (1978), see section 9.12, with 10 initial points.

3. The adaptive Bayesian by Mockus (1984), see section 9.9, with 25 initial points and 500 observations.

4. The extrapolation algorithm of Zilinskas (1980), see section 9.10, with parameters LT = 30 and ML = 3.

Standard No.	f_0 is opt. value / N is no. of obs.	Algorithm No. 1		Algorithm No. 2		Algorithm No. 3		Algorithm No. 4	
		Objective No. 1	Objective No. 2	Objective No. 1	Objective No. 2	Objective No. 1	Objective No. 2	Objective No. 1	Objective No. 2
1	f_0	0.048	1.34	0.0046	4.79	0.019	1.07	0.0076	0.0006
	N	3596	3439	6298	9042	1150	1934	2765	3128
2	f_0	7.54	11.8	0.051	11.8	0.051	11.8	0.051	11.8
	N	3508	3328	4330	8007	1065	910	3770	1966
3	f_0	0.032	7.54	0.035	2.78	0.026	4.21	0.017	0.00004
	N	3651	3508	6462	9826	1145	1324	4075	3839

Table 8.9.1

Results of minimization of spectral and colour differences by four global algorithms for three standard samples and a set of nine pigments

The result of global optimization was corrected using the version of the variable metrics algorithm by Tieshis (1975), see section 9.16, with the following parameters:

initial step of numerical differentiation 10^{-4}
tolerance of the step length 10^{-3}
tolerance of the norm of gradient 10^{-4}
tolerance of the decreasing function during one iteration 10^{-4}.

Table 8.9.1 shows that methods 3 and 4 are clearly more efficient in the sense that they provide better results than methods 1 and 2 with less observations. Comparison of methods 3 and 4 is not so easy because method 4 gives better function values than method 3, but method 3 needs less observations than method 4.

To arrange an equal number of observations for both methods is difficult because method 4 stops by its own stopping rule and the number of observations of method 3 is fixed.

It is possible to force method 4 to stop at the fixed number of observations. However, in this case, the efficiency of the method can be reduced since a good stopping rule is one of the advantages of method 4.

Figures 8.9.1 and 8.9.2 show the goodness of fit of the optimized (broken lines) spectra to the standard spectra 1 and 2, respectively.

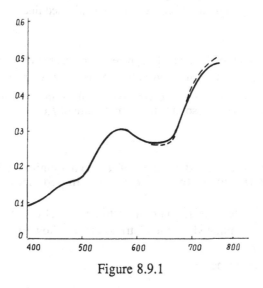

Figure 8.9.1

Spectra of the standard (continuous line) and the composition corresponding to minimal mean square deviation (dotted line)

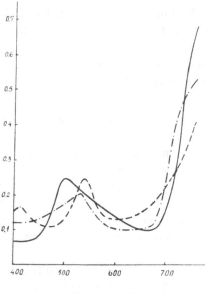

Figure 8.9.2

Spectra of the standard No. 2 (continuous line) and of the compositions
corresponding to minimal mean square deviation (broken line with dots) and to the
minimal difference of colours (dashed line)

The worst fit was in case No.2 which represents an organic dye. However, even in
this case the difference appeared, visually, to be only a tint of the same colour.
The broken line with dots (—·—·—) corresponds to the minimum of function $f_1(x)$.
The dashed line (– – –) corresponds to the minimum of $f_2(x)$.

8.10   The least square estimation of electrochemical adsorption using observations of the magnitude of electrode impedance

The process of inhibited adsorption of an active material in the absence of density
restrictions can be simulated by the elctrical circuit shown in Figure 8.10.1, see
Dzemyda et al (1984).

Denote the impedance of the circuit as

$$Z = R_1 + \left[j\omega C_1 + (R_2 + 1/(j\omega C_2))^{-1} \right]^{-1}$$

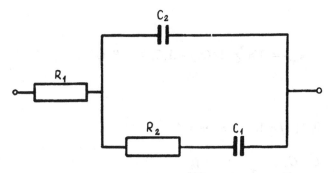

Figure 8.10.1

The electrical circuit

To make the observations of electrode parameters more accurate the resistance R_1 is balanced. Then the observed impedance is the difference $Z_b = Z - R_b$, where R_b is the balancing resistance.

We shall express the magnitude of the impedance as

$$|Z_b| = A(C, K, \gamma, R, \omega)$$

where

$$
\begin{array}{ll}
R = R_1 - R_b, & R_a \leq R \leq R_d, \\
C = C_1 + C_2, & C_a \leq C \leq C_d, \\
K = R_2 C_2, & K_a \leq K \leq K_d, \\
\gamma = C_1/(C_1 + C_2), & \gamma_a \leq \gamma \leq \gamma_d, \\
\omega = 2\pi_0 f,
\end{array}
$$

and f_0 is the frequency; the index a denotes the lower limit and the index d indicates the upper limit of the correponding parameter.

The empirical values of the magnitude of impedance at fixed ω_i are denoted by $A(\omega_i)$. Then the least square estimate of the parameters can be defined by the following condition

$$\min_{x_1,\dots,x_m} \ f(x_1, \dots, x_m)$$

where

$$f(x_1, \ldots, x_m) = 1/n \sum_{i=1}^{n} [A(\omega_i) - A(C, K, \gamma, R, \omega)]^2$$

and

$$x_j \in [0, 1], \quad j = 1, \ldots, m, \quad m = 4,$$

$$x_1 = \frac{C - C_a}{C_d - C_a}, \quad x_3 = \frac{\gamma - \gamma_a}{\gamma_d - \gamma_a},$$

$$x_2 = \frac{K - K_a}{K_d - K_a}, \quad x_4 = \frac{R - R_a}{R_d - R_a}.$$

The global optimization was done by the uniform deterministic method of LP type by Sobolj (1969), Dzemyda (1983) and Shaltenis and Dzemyda (1982) with the ordering of variables by the methods of Dzemyda (1982), using the routine LPTAU, see secton 9.11. The number of observations for the analysis of structure was 40 and for the LP optimization was 300.

The local optimization was carried out by Barauskas version of the variable metrics method, see Barauskas (1984) with parameters as recommended by the author.

First to estimate the error of optimization, the circuit in Figure 8.10.1 with fixed parameters was considered. Table 8.10.1 shows the difference between the actual values of parameters and their estimates.

Parameter	actual value	estimated value	error of estimation %
$C, \mu F$	20.90	21.60	3.2
K	140.50	142.00	1.0
γ	0.048	0.044	8.3
R, Ω	-4.53	-4.41	2.6

Table 8.10.1

Difference between actual and estimated values of the parameters of the electrical circuit shown in Figure 8.10.1

The maximal error was 8.3% which was considered acceptable and therefore the method was used in the real problem. The least square estimates of the parameters C, K, γ, R which define the adsorption of atomic hydrogen on the platinum electrode in solutions 1 N H_2SO_4 and 1 N $H_2SO_4 + 2.5.10^{-3}$ N Zn^{2+} are shown in Table 8.10.2. Since the adsorption of Zn^{2+} is slow the process was approximately described by the electrical circuit shown in Figure 8.10.1. (See Joshida et al (1978).)

Solution	E, V	Parameters			
		$C, \mu F$	K	γ	R, Ω
1 N H_2SO_4	$E_p = 0.12$	671	280	0.02	– 0.2
	$E_m = 0.19$	315	180	0.02	– 0.2
1 N H_2SO_4 + $2.5.10^{-3}$ N Zn^{2+}	$E_p = 0.16$	404	240	0.05	– 0.2
	$E_m = 0.21$	349	240	0.05	– 0.2

Table 8.10.2

Estimated parameters of adsorption for two solutions

In Table 8.10.2 the potential E_m means the beginning of the adsorption of H_m and potential E_p defines the peak of the internal transition $H_1 \Leftrightarrow H_m$, see Hochshtein (1976).

The results for the solution 1 N H_2SO_4 correspond to similar results by Hochshtein (1976). The only difference is that K^{-1} in Table 8.10.2 was smaller. In the presence of Zn^{2+} the specific potential of adsorption is biased towards the anode; the capacity, which corresponds to the peak of internal transition, is much smaller and the resistance R is greater.

8.11 Estimation of parameters of the immunological model

The simplest mathematical model of an immune response suggested by Marchuk in 1975, see Marchuk (1980), is the system of nonlinear differential equations

$$dV/dt = -\gamma F V$$
$$dF/dt = \rho C - \eta \gamma F V - \mu_1 F$$
$$dC/dt = \alpha F V|_{t-\tau} - \mu_2(C - C_0)$$
$$V(0) = V_0, \ F(0) = F_0 = \rho C_0 | \mu_1, \ C(0) = C_0 \qquad\qquad (8.11.1)$$

where

$V = V(t)$ is the the density of antigen,

$F = F(t)$ is the density of specific antibodies,

$C = C(t)$ is the density of plasma cells of mature antibody producers,

$\gamma, \rho, \eta, \mu_1, \mu_2, \alpha$ are the unknown parameters of the model, and

τ is the delay time.

Zuev (1982) suggested estimating the unknown parameters by maximization of the likelihood function using the experimental data. The deterministic model (8.11.1) was extended to the stochastic one, so that the maximum likelihood method could be used. The stochastic model of Zuev involves some additional parameters so that the total number of parameters to be estimated is nine; three parameters of the stochastic part and six parameters of the deterministic part, similar to (8.11.1). The relation between the parameters of the model and the corresponding values of $V(t)$, $F(t)$ and $C(t)$ at the points t_i, $i = 1, \dots , K$ are defined using numerical methods for the integration of systems of ordinary differential equations with constant time delay developed by Belykh (1982).

The likelihood function was not convex so the global Bayesian method, see section 9.9, was used to locate the vicinity of the minimum. In Figure 8.11.1 the broken line shows how the expected density of plasma cells and its variance depend on time.

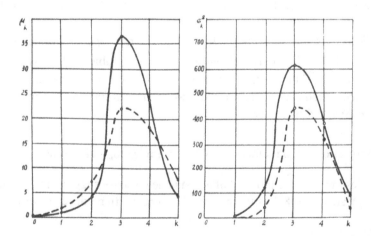

Figure 8.11.1

Mean value μ_k and variance σ_k^2 of plasma cell density during immunological reaction

The continuous line shows the experimental results (average from 10 samples). The experiments were performed by Asachenkov and Stepanenko. The experimental results correspond to the reaction of a homogeneous sample of mice to the inoculation of a nonpathogenic antigen.

8.12 The optimization of nonstationary queuing systems

The simple version of the nonstationary queuing system is considered when the flow of events is deterministic, with the density of events

$$a(t) = \begin{cases} a_1, & 0 \leq t \leq t_0 \\ a_2, & t_0 < t \leq T \end{cases} \tag{8.12.1}$$

and the service time is uniformly distributed in the intervals $[0, 1/x_1]$ and $[0, 1/x_2]$ depending on t, hence the density of the service time

$$b(t) = \begin{cases} x^1, & 0 \leq t \leq x_3 \\ x^2, & x^3 \leq t \leq T \end{cases} \tag{8.12.2}$$

The objective function is assumed to be the sum

$$f(x) = \sum_i q_i + x^1 x^3 + x^2 (T - x^3) \tag{8.12.3}$$

where $\sum_i q_i$ is the queuing cost, q_i is a queuing time, x^1, x^2 are the operational costs per time unit.

The last column of Table 8.12.1 shows the difference between the results of the corresponding method and the optimal decision. The differences are calculated using the following formula

$$\delta = |x_N^1 - 2.30| + |x_N^2 - 3.50| + \min_{0 \leq v \leq 4} |x_N^3 - 5 - v| \tag{8.12.4}$$

The last component of formula (8.12.4) means that the variable x^3 is nearly optimal in the interval [5.0, 9.0]. In (8.12.4) the deviation is calculated using the absolute values of differences with unit weights, which is assumed to be natural in the circumstances.

No.	Method	x_0			Start point	x_N			N	T	δ
		x^1	x^2	x^3		x^1	x^2	x^3			
1	2	3	4	5	6	7	8	9	10	11	12
1	BAYES1					3.23	8.39	9.56	1000	179.4	5.82
2	LBAYES	3.23	8.39	5.56	From no. 1	2.40	3.70	8.31	11699	3.1	0.30
3	BAYES1					2.66	3.51	8.60	11738	2.37	4.01
4	LBAYES	2.66	6.70	4.52	From no. 3	2.39	3.51	8.60	11738	2.56	0.10
5	LBAYES	7.20	6.30	2.94	Random	2.09	3.57	5.10	11699	2.51	0.28
6	--	5.25	1.75	8.93	--	2.34	3.70	7.53	--	--	0.24
7	--	2.66	9.24	4.11	--	2.10	3.51	5.02	--	--	0.21
8	--	4.80	9.65	6.01	--	2.49	3.92	9.90	--	--	0.61
9	--	8.92	6.69	8.91	--	2.18	2.72	0.08	--	--	5.0
10	--	6.53	2.48	1.34	--	2.05	2.89	1.42	--	--	4.34
11	--	7.91	1.15	2.76	--	5.88	2.79	0.10	--	--	9.19
12	--	4.97	4.84	7.25	--	2.39	2.71	0.15	--	--	5.63
13	--	5.33	3.09	8.67	--	2.19	3.64	5.29	--	--	0.24
14	--	5.88	2.79	0.10	From no. 11	2.20	2.72	0.63	--	--	6.11
15	--	2.20	2.72	6.31	From no. 14	1.91	2.94	2.21	--	--	3.74
16	--	1.91	2.94	2.21	From no. 15	2.26	3.48	5.67	--	--	0.06

Table 8.12.1

Results of optimization of a queuing system

Three types of method were used. The first type was the sequence of global and local search, see Nos. 1and 2 and Nos. 3 and 4. In the case of No. 1 BAYES1 was used with the number of uniform observations LT = 5; in case No. 3 the same global method was used with LT = M = 1000 which means the uniform LP-search. In cases 1 and 2 the final deviation was 0.30, in cases 3 and 4 it was 0.10. As the second type, we considered local search with randomly distributed starting points, see Nos. 5 to 13. The minimal error from all cases was 0.21, the average deviation was 2.86.

The third type was the local search from a random point represented as a sequence of three local methods (Nos. 14 to 16).

The best deviation was 0.06. Column 11 shows the CPU time for the BESM-6 computer.

We can see that the function is obviously unimodal. In terms of the best result the deviations of global plus local search (corresponding to Nos. 1 and 2 and Nos. 3 and 4) are similar to the final deviation of the sequence of three local search procedures (Nos. 14 to 16), or the best of local search procedure repeated nine times (Nos. 5 to 13).

In terms of numbers of observations the global-local approach (Nos. 1 and 2 and Nos. 3 and 4) was almost three times more economical when compared with the local-local approach (Nos. 14 to 16).

In terms of CPU time the global-local search NOs. 3 and 4 was best. In this case the global-local method corresponded to uniform LP-search.

8.13   The analysis of structure of the Steiner problem

The Steiner problem is how to connect a given set of $N \geq 3$ points by a graph of minimal length. The Steiner problem is combinatorial because there exists a finite number of possible ways of connecting the fixed points using additional points. This problem can be reduced to the continuous but multimodal problem of how to define such co-ordinates of $n \leq N - 2$ additional points which minimize the sum of the length of arcs of the minimal connecting graph, see Shaltenis (1976). Such a graph can easily be found using the well known algorithm of Primm if the co-ordinates of all points are fixed. The difficult problem is to define the co-ordinates of $n \leq N - 2$ additional points, when the number N of fixed points is large.

The algorithm of Shaltenis (1975) was used to minimize the length of the network. The idea of the algorithm is similar to that of the Bayesian algorithm.

The main difference between the Shaltenis algorithm and the regular Bayesian one as defined by condition (5.1.1) can be described in the following way. Algorithm (5.1.1), at each   step, defines a point of the next observation which minimizes the risk function (5.1.1). The Shaltenis algorithm defines not the next point of observation but the next 'strategy' of search which minimizes the risk function.

There are $n + 1$ strategies available: one is the n-dimensional Monte Carlo search, the other n strategies correspond to one-dimensional uniform random search along each co-ordinate.

The conditional expectations and conditional variances are defined approximately by some simple formulae, using the results of observations. The choice of the descibed algorithm can be explained in two ways:

1. Observations are not expensive, so the simple method should be used.

2. Interdependence of variables is not expected to be be very strong, so the co-ordinate optimization is reasonable.

The results are shown in Figures 8.13.1 and 8.13.2. The continuous line in Figure 8.13.1 corresponds to the result of the algorithm after 1225 observations. The dotted line corresponds to the optimal decision. Figure 8.13.1 shows that the algorithm found the decision which is near to the global minimum. To fix the exact minimum it is better to use some procedure of local optimization rather than to try to do it by increasing the number of observations of the global method.

Figure 8.13.1

Result of minimization of connecting graph

Figure 8.13.2 shows how the length of the network depends on the number of observations. The figure corresponds to the Steiner problem with 12 fixed points (20 variables). The dotted lines correspond to the usual Monte Carlo procedure, see section 9.13.

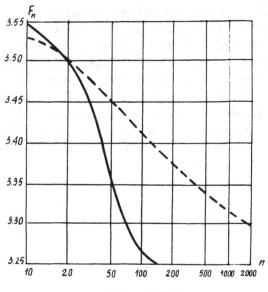

Figure 8.13.2

The relationship of the length of the minimized graph to the number of observations

We can see that the comparative efficiency of the Shaltenis (1975) algorithm increases with the dimensionality of the problem and with the number of observations.

8.14 The estimation of decision making by intuition on the example of the Steiner problem

The Steiner problem is a convenient test problem to estimate the errors of decisions made by intuition. The estimation is interesting because many of the decisions in the design of large engineering systems are made without any formal optimization procedures. So it is important to see how muchfcan be gained when the regular optimization methods are applied. A good point of the Steiner problem is that it is easy to understand and difficult to solve because the number of local minima is very large. For example, in the case of ten fixed points, the number of local minima is more than 2.10^6.

The usual regression analysis was done using the results of 1500 test problems under different conditions, see Mockus (1967). As expected it shows a positive relation between the mean error and such factors as the complexity of the test problem (number of fixed points), and negative relation of the mean error and the time which was available to make the decision, the level of education (estimated by years of education) and the academic grades (in the five grade system).

Unexpectedly it was found that the mean error monotonically depends on the level of experience (the number of Steiner problems already solved).

The minimal error corresponds to the third problem in the sequence of eight Steiner problems. The mean error decreased sharply until the third problem and then increased slowly. The most obvious explanation of this phenomenon is the decreasing interest after the consideration of three similar test problems.

The general average was 12% and the minimal error of the best intuitive decision was 0.2%. This is rather surprising since the problem is very difficult indeed.

CHAPTER 9

PORTABLE FORTRAN SOFTWARE FOR GLOBAL OPTIMIZATION

9.1 Introduction

The purpose is to minimize a continuous function

$$f(x), \quad x = (x_1, \dots, x_n) \tag{9.1.1}$$

where $x \in A \subset R^n$.

It is assumed for most methods that the set A is a rectangular parallelepiped

$$A = \{x : a_i \le x_i \le b_i, \ i = 1, \dots, n\} \tag{9.1.2}$$

because the more general case can be approximately reduced to (9.1.2) using the penalty function techniques. The general idea is to implement the methods which can be regarded as optimal in the sense of average deviation. In defining the average optimality the number of observations is usually taken into account. The procedures which minimize the risk function should be as simple as possible but some natural limits exist. For example, in the case of global optimization of continuous functions, in order to minimize the average deviation the risk function should be minimized. It cannot be reduced to the unimodal one if the convergence to a global minimum of any continuous function is to be provided. So it seems that the global Bayesian method described by Mockus (1984[a]) is possibly the simplest one which minimizes the average deviation and converges to a minimum. If the function $f(x)$ is simple enough then it can be that a better policy is not to minimize the risk function at all but to take more observations, which need not be planned optimally.

In such a case it is a good idea to use uniform search, for example the LP type, see Sobolj (1969). In this case we lose the average optimality but still have the convergence. Even the LP-search can be too expensive if the observations are very cheap. Then we can use the uniform random Monte Carlo search which is very simple indeed, but in this case we shall have convergence only in the probabilistic sense. For these reasons both the LP search and the uniform random search are included in this package. The method of clustering by Törn (1978) can be regarded as something between the LP search and Monte Carlo.

197

For local optimization of continuously differentiable functions without noise the methods of variable metrics type seem to be the best. Different versions of those methods are included: one is for rectangular constraints, the other is for nonlinear constraints.

For local optimization of nondifferentiable functions with nonlinear constraints the simplex type method is used, see Himmelblau (1972). If there are some reasons to expect that the influence of some variables and their groups is considerably greater than that of others then the method of analysis of structure should be used before the usual optimization, see Shaltenis and Dzemyda (1982).

For the local optimization of unimodal functions with noise a good compromise between simplicity and optimality is the stochastic approximation type of methods with Bayesian step length, see Mockus (1984[b]).

By adjusting machine dependant constants, the software can be adapted to any computer with a standard FORTRAN compiler. No machine dependent routines are used.

9.2   Parameters

The parameters X, A, B, N, FM, IPAR, PAR, IPA, IPAA are used in all subroutines.

X is an array of length N which defines the co-ordinates of a point being considered (initial, optimal or current).

A is an array of length N which defines the lower bounds of X.

B is an array of length N which defines the upper bounds of X

N is the number of variables (dimension of X) usually $N \leq 20$.

FM is the value of function FI(X, N) at the point X.

IPAR is the array of length 30 which defines the integer control parameters.

PAR is an array of length 30 which defines the real control parameters.

IPA is the shift of integer control parameters.

IPAA is the shift of real control parameters.

If only one method is used then both shifts are zero: IPA = 0 and IPAA = 0.

If several methods are used sequentially then the shift for the next method must be equal to the sum of numbers of control parameters used before by other methods. The number of control parameters of different methods are given in Table 9.3.1. For all the methods the first integer control parameter is the printing parameter:

IPAR(IPA + 1) = IPR   (IPA = 0 if only one method is used)

If IPR < 0

then only diagnostic messages are printed,

if IPR = 0

then the initial data, the final results and the diagnostic messages are printed,

if IPR > 0

then not only those but also the results of each IPR-th iteration are printed.

The meaning of other control parameters will be explained later when describing corresponding subroutines.

9.3   Methods available

No.	Name	Method	No.of control parameters	
			Integer IPAR	Real PAR

Global optimization with rectangular constraints

No.	Name	Method	Integer IPAR	Real PAR
1	BAYES1	Bayesian, see Mockus (1984[a])	3	0
2	UNT	Extrapolation, see Zilinskas (1982)	4	0
3	LPMIN	Uniform deterministic, see Shaltenis and Dzemyda (1982) and Sobolj (1969)	N+3	0
4	GLOPT	Clustering, see Törn (1978)	3	0
5	MIGI, MIG2	Uniform random	2	0
6	EXTR	Bayesian one-dimensional, see Zilinskas (1978[b])	3	2

Local optimization

7	MIVAR4	Variable metrics with rectangular constraints, see Tieshis (1975)	4	4
8	REQP	Variable metrics with nonlinear constraints, see Biggs (1974)	4	4
9	FLEXI	Simple with nonlinear constraints, see Himmelblau (1972)	4	2

Local optimization with noise

10	LBAYES	Stochastic approximation with Bayesian step length and rectangular constraints, see Mockus (1984[b])	3	2

Analysis   of structure

11	ANAL1	Estimation of influence of variables and their pairs, see Shaltenis and Radvilavichiute (1976)	5	0

Table 9.3.1

9.4   Common blocks

/BS1/ Y(1000) are the values of function FI(X, N) at the points form the array XN of length MN = N * M which contains co-ordinates of M points of function evaluations

/STATIS/ IFAIL, IT, IM, M

Here

IFAIL is the control indicator if the initial data is not correct

then

IFAIL = 10 and return to the main program,

if the initial data is correct then IFAIL ≠ 10 and shows the number of the termination criterion

IT is the number of iterations,

IM is the number of the optional iteration (where the best point was found),

M is the number of function evaluations (observations).

9.5   The function

The function to be minimized should be represented as real function FI(X, N)

In most methods only the lower and upper bounds should be fixed by arrays A and B

In methods with nonlinear constraints the subroutine

CONSTR (X, N, G, MC)

should be used, where

G is a one-dimensional array of length MC which defines the constraints at the point X

MC is the number of constraints

It is well known that the local methods of optimization generally are sensitive to the scales of variables. The parameters of local methods in this package usually are adjustedto the case when $A(I) = -1$ and $B(I) = 1, I = 1, N$. So it can be useful to reduce the rectangular constraints $A \leq X \leq B$ to the N-dimensional hypercube $[-1, 1]^N$, which can be arranged using the reduction formulae (in the case of LBAYES those formulae are included in the algorithm, so here no additional reduction is needed).

The reduction formulae are as follows

$$X(I) = \frac{2}{B(I) - A(I)} \, X0(I) - \frac{B(I) + A(I)}{B(I) - A(I)} \, , \, I = 1, N$$

where $X0$ are the original variables and X are the variables scaled to fit into $[-1, 1]^N$.

EXAMPLE. In most of the following examples of methods, the following test function $f(x)$ is used

$$f(x) = 2/N \sum_{i=1}^{N} \left(x_i^2 - \cos\left(18x_i\right) \right) \tag{9.5.1}$$

with $N = 2$, $x_1 \in [-0.25; 0.5]$, $x_2 \in [-0.125; 0.625]$

which is represented by subroutine function FURASN:

```
FUNCTION FURASN (X, N)

DIMENSION X(N)
      F=0
      DO 10 I = 1, N
         XI=X(I)
10    F=F + XI * XI – cos(18 * XI)
         FURASN = F*(2./FLOAT(N))
         RETURN
         END
```

9.6 The main program

The main program defines the input data and the sequence of methods. At the beginning we should choose, from Table 9.3.1, the desirable sequence of methods of optimization and analysis. Then the function FI(X, N) is prepared which evaluates the objective function at a fixed point X. If necessary the subroutine CONSTR (X, N, G, MC) is included which evaluates the constraints at the point X. The length of arrays usually depends on the subroutines which are used. The length of arrays IPAR and PAR is always 30. The parameters of methods are included in the arrays IPAR, PAR in accordance with the given sequence of methods. The formal parameters IPA, IPAA are fixed using the rules given in the previous section. In the case when only one method is used IPA = IPAA = 0.

9.7 The example of the main program

The following test problem is considered. The global minimum of test function (9.5.1) should be defined using the global Bayesian method BAYES1, then the results of global search should be corrected by the local method of variable metrics MIVAR4. The test function is represented as the function FURASN (X, N).

The arrays are: X, A, B, XN, HES, IPAR, PAR.

It follows from Table 9.3.1 that in the subroutine BAYES1 three integer parameters, and in subroutine MIVAR4, four integer parameters are used. In MIVAR4 four real parameters should be defined. This means that seven elements of IPAR and four elements of PAR should be fixed.

The main program:

```
DIMENSION X(2), A(2), B(2), XN(200), HES(3), IPAR(30), PAR(30)
DATA N, NM, NH, IPA, IPAA/2, 200, 3, 0, 0/, A/ –0.25, – 0.125/, B/0.5,
0.625/
DATA IPAR/0,100, 0, 100, 2, 100, 23*0/, PAR/100., 3*1.E–4, 26*0./
CALL BAYES1 (X, A, B, N, XN, NM, FM, IPAR, IPA)
IPA = 3
CALL MIVAR4 (X, A, B, N, HES, NH, FM, IPAR, PAR, IPA, IPAA)
STOP
END
FUNCTION FI(X, N)
DIMENSION X(N)
FI=FURASN (X, N)
RETURN
END
```

9.8 Description of routines

In this chapter the general description of routines corresponding to different methods of global and local optimization will be given. The general description of routines includes the name and purpose of the routine, the restrictions and the accuracy. It also shows how to use the routine. A simple example is provided.

The FORTRAN codes are given separately in the diskette, because the complete listings are needed only for advanced users who want to check or to change the parameters and procedures of optimization. In the general description of algorithms and programs only the most important information is provided.

9.9   BAYES1: the global Bayesian method by Mockus

PURPOSE. To find the global minimum of a continuous function (9.1.1) of N variables defined on the rectangular parallelepiped (9.1.2).

RESTRICTIONS. In terms of efficiency this program becomes increasingly less successful as the dimensions of the parallelepiped increase. So the dimensions must be as small as possible

The global Bayesian method uses a considerable amount of auxiliary calculation.   As a result no more than 1000 function evaluations can be performed. This means that BAYES1 can be efficiently used only when the function $f(x)$ is difficult – takes more CPU time, than the calculation of the co-ordinates of the next point. The global method does the search in the whole area. It can find the point in the neighbourhood of a global minimum but cannot always fix the point of minimum with sufficient accuracy. So some local methods should be used to carry out the local minimization.

ACCURACY. The global method provides the minimal average deviation in accordance with a given statistical model, see Mockus (1984[a]) and the convergence to a global minimum for any continuous function. This means that if we solve many problems the average error will be as small as possible. However, for some fixed samples it can be large if the iteration number is limited.

HOW TO USE THE METHOD

CALL BAYES1 (X, A, B, N, XN, NM, FM, IPAR, IPA)

where the input is: A, B, N, NM, IPAR, IPA

and the output is X, XN, FM.

XN is an array of length NM = N * M which contains co-ordinates of M points of function evaluation.

In the main program the following arrays should be described:

A(N), B(N), X(N), XN(NM), IPAR(30)
$N \leq 20$
IPAR(1) = IPR is printing parameter
IPAR(2) = M is number of function evaluation $M \leq 1000$.
IPAR(3) = LT is number of initial points which are uniformly distributed by the LP sequences of Sobolj (1969)
$0 < LT \leq M$.

EXAMPLE. The program locates the minimum of the multimodal function (9.5.1).

IPR = 0, M = 100, LT = 5. The initial information is fixed by DATA.

```
DIMENSION X(2), A(2), B(2), XN(200), IPAR(30),
DATA N, NM,  IPA/2, 200, 0/, A/ – 0.25,  – 0.125/, B/0.5, 0.625/
DATA IPAR/0, 100, 5, 27*0/
CALL BAYES1 (X, A, B, N, XN, NM, FM, IPAR, IPA)
 STOP
END

FUNCTION FI(X, N)
DIMENSION X(N)
FI=FURASN (X, N)
RETURN
END
```

BAYES1

INITIAL DATA

NUMBER OF VARIABLES $N = 2$
PRINTING PARAMETER $IPR = 0$
NUMBER OF FUNCTION EVALUATIONS $M = 100$
NUMBER OF INITIAL POINTS $LT = 5$

VECTOR OF LOWER BOUNDS (A) FOR X
$-0.2500000E\ 00\ \ -0.12500000E\ 00$
VECTOR OF UPPER BOUNDS (B) FOR X
$0.5000000E\ 00\ \ \ 0.62500000E\ 00$

– –
– –

RESULTS

OPTIMAL FUNCTION VALUE FM = $-0.19982195E\ 01$ OBTAINED IN $NR = 45$
OPTIMAL POINT
$0.10454655E{-}02\ \ -0.31354427E{-}02$
NUMBER OF FUNCTION EVALUATIONS $L = 100$

BAYES1 TERMINATED

– –
– –

9.10 UNT : The global method of extrapolation type by Zilinskas

PURPOSE. To find the global minimum of a continuous function (9.1.1) of N variables defined on the rectangular parallelepiped (9.1.2).

RESTRICTIONS. The restrictions are the same as in the BAYES1 method except that if the function is differentiable then the local search of variable metrics type can be directly incorporated in UNT.

ACCURACY. The method provides the minimal average deviation in accordance with the set of assumptions given by Zilinskas (1982). All other accuracy considerations are similar to those in the BAYES1 method.

HOW TO USE THE METHOD.

CALL UNT (X, A, B, N, XN, NM, FM, IPAR, IPA)

where the input is: A, B, N, NM, IPAR, IPA

and the output is X, XN, FM.

XN is an array of length NM = N * M which contains co-ordinates of M points of function evaluation.

In the main program the following arrays should be described:

A(N), B(N), X(N), XN(NM), IPAR(30)
$N \le 20$
IPAR(1) = IPR is printing parameter
IPAR(2) = M is the maximal number of function evaluation, $M \le 500$.
IPAR(3) = LT is number of RANDOM uniformly distributed initial points, $LT \ge 30$
IPAR(4) = ML is the maximal number of local minima, $0 < ML \le 20$

EXAMPLE. The program locates the global minimum of the multimodal function (9.5.1).

IPR = 0, M = 500, LT = 30, ML = 5.

```
DIMENSION X(2), A(2), B(2), XN(1000), IPAR(30),
DATA N, NM, IPA/2, 1000, 0/, A/– 0.25, – 0.125/, B/0.5, 0.625/
DATA IPAR/0,500, 30, 5, 26*0/
```

```
CALL UNT (X, A, B, N, XN, NM, FM, IPAR, IPA)
STOP
END

FUNCTION FI(X, N)
DIMENSION X(N)
FI=FURASN (X, N)
RETURN
END
```

UNT

INITIAL DATA

NUMBER OF VARIABLES	N = 2
PRINTING PARAMETER	IPR = 0
MAXIMUM NUMBER OF FUNCTION EVALUATIONS	M = 500
NUMBER OF INITIAL POINTS	LT = 30
MAXIMUM NUMBER OF LOCAL MINIMA	ML = 5

VECTOR OF LOWER BOUNDS (A) FOR X
– 0.2500000E 00   – 0.12500000E 00
VECTOR OF UPPER BOUNDS (B) FOR X
0.5000000E 00   0.62500000E 00

– –
– –

RESULTS

OPTIMAL FUNCTION VALUE FM = – 0.19621372E 01
OPTIMAL POINT
0.15230108E–021   0.13288877E–02
      LOCAL OPTIMA

FUNCTION VALUE	POINT	
– 0.19621372E 01	0.15230108E–01	0.13288877E–02
– 0.18478394E 01	– 0.13205975E–01	0.34279919E 00
– 0.18367290E 01	0.34698546E 00	0.16140953E–01
–0.17412872E 01	0.34251302E 00	0.33785129E 00

NUMBER OF FUNCTION EVALUATIONS   L = 142

IFAIL = 1. TERMINATION CRITERION: DENSITY OF POINTS IN LOCAL
MINIMUM AREA EQUALS 3

UNT TERMINATED

9.11   LPMIN: The global method of uniform search by Sobolj, Shaltenis and Dzemyda

PURPOSE.   To locate the global minimum of a continuous function (9.1.1) of N variables defined on the rectangular parallelepiped (9.1.2).

RESTRICTIONS.   The restrictions are the similar to those in the BAYES1 method with the exception that LPMIN is using less auxilliary calculations and therefore it can be recommended for minimising simpler functions than BAYES1 or UNT.

ACCURACY.   The method provides convergence to a minimum.   The average deviation is usually considerably greater than in the methods BAYES1 and UNT.

HOW TO USE THE METHOD.

CALL LPMIN (X, A, B, N, XN, NM, FM, IPAR, IPA)

where the input is: A, B, N, NM, IPAR, IPA

and the output is X, XN, FM.

XN is an array of length NM = N * M which contains co-ordinates of M points of function evaluation.

In the main program the following arrays should be described:

A(N), B(N), X(N), XN(NM), IPAR(30)
N ≤ 20

IPAR(1) = IPR is printing parameter

IPAR(2) = M is the indicator of analysis of structure. IF M < 0 then no analysis of structure is performed. If $10 \leq M \leq 300$ then the results of M observations are used to number the variables in order of decreasing importance. If M = 0 then the order of variables should be fixed by the user in accordance with his opinion about the decreasing importance

IPAR(3) = ML is number of steps of LP search

IPAR(4)
$---$ are the number of variables in order of decreasing
$---$ importance which should be fixed when IPAR(2) = M = 0
IPAR(N+3)

EXAMPLE. The program locates the global minimum and performs the analysis of structure of the multimodal function (9.5.1).

IPR = 0, M = 50, ML = 1000.

```
DIMENSION X(2), A(2), B(2), XN(100),  IPAR(30),
DATA N, NM,  IPA/2, 100, 0/, A/ – 0.25,  – 0.125/, B/0.5, 0.625/
DATA  IPAR/0, 50, 1000, 27*0/
CALL LPMIN (X, A, B, N, XN, NM, FM, IPAR, IPA)
STOP
END

FUNCTION FI(X, N)
DIMENSION X(N)
FI=FURASN (X, N)
RETURN
END
```

LPMIN

INITIAL DATA

NUMBER OF VARIABLES	N = 2
PRINTING PARAMETER	IPR = 0
NUMBER OF FUNCTION EVALUATIONS FOR ANALYSIS	M = 50
NUMBER OF FUNCTION EVALUATIONS FOR LP-SEARCH	ML = 1000

VECTOR OF LOWER BOUNDS (A) FOR X
$-0.2500000E\ 00\ \ -0.12500000E\ 00$
VECTOR OF UPPER BOUNDS (B) FOR X
$0.5000000E\ 00\ \ \ 0.62500000E\ 00$

LP-SEARCH WITH ANALYSIS
– –
– –
RESULTS OF ANALYSIS

VARIABLES BY DECREASING INFLUENCE
     2   1

OPTIMAL FUNCTION VALUE FM = $-0.19209299E\ 01$ OBTAINED IN
NR = 10
OPTIMAL POINT
$-0.15625000E{-}01\ \ \ 0.15625000E{-}01$
– –
– –
RESULTS OF LP-SEARCH

OPTIMAL FUNCTION VALUE FM = $-0.19915352E\ 01$ OBTAINED IN
NR = 404
OPTIMAL POINT
$O.63476563E{-}02\ \ \ -0.34179688E{-}02$
– –
– –
RESULTS

OPTIMAL FUNCTION VALUE FM = $-0.19915352E\ 01$
OPTIMAL POINT
$O.63476563E{-}02\ \ \ -0.34179688E{-}02$

NUMBER OF FUNCTION EVALUATIONS   L = 1050

LPMIN TERMINATED

– –
– –

9.12 GLOPT: The global method of clustering type by Törn

PURPOSE. To find the global minimum of a continuous function (9.1.1) of N variables defined on the rectangular parallelepiped (9.1.2).

RESTRICTIONS. The restrictions are the same as in the methods BAYES1, UNT, LPMIN with the exception that GLOPT uses a smaller number of auxilliary calculations so it can be recommended for minimising simpler functions when compared with all previous methods if convergence is not necessary.

ACCURACY. The convergence to a minimum is not provided but the accuracy usually satisfies the practical needs.

HOW TO USE THE METHOD.

CALL GLOPT (X, A, B, N, FM, IPAR, IPA)

where the input is: A, B, N, IPAR, IPA

and the output is X, FM.

In the main program the following arrays should be described:

A(N), B(N), X(N), IPAR(30)
N ≤ 20
IPAR(1) = IPR is printing parameter
IPAR(2) = M is the maximal number of function evaluations, recommended M = 10000
IPAR(3) = LT is number of random uniformly distributed initial points, LT ≤ 150, recommended LT is about double the number of an expected number of local minima.

EXAMPLE. The program locates the global minimum of the multimodal function (9.5.1).

IPR = 0, M = 10000, LT = 10.

```
      DIMENSION X(2), A(2), B(2), IPAR(30),
      DATA N,  IPA/2, 0/, A/ – 0.25,  – 0.125/, B/0.5, 0.625/
      DATA IPAR/0,10000, 10, 27*0/
      CALL GLOPT (X, A, B, N, FM, IPAR, IPA)
      STOP
```

    END

    FUNCTION FI(X, N)
    DIMENSION X(N)
    FI=FURASN (X, N)
    RETURN
    END

GLOPT

INITIAL DATA

NUMBER OF VARIABLES   N = 2
PRINTING PARAMETER   IPR = 0
MAXIMUM NUMBER OF FUNCTION EVALUATIONS   M = 10000
NUMBER OF INITIAL POINTS   LT = 10

VECTOR OF LOWER BOUNDS (A) FOR X
– 0.2500000E 00   – 0.12500000E 00
VECTOR OF UPPER BOUNDS (B) FOR X
0.5000000E 00   0.62500000E 00

– –
– –

RESULTS

OPTIMAL FUNCTION VALUE FM = – 0.19999999E 01
OPTIMAL POINT
– 033581266E–04   0.37407226E–04

NUMBER OF FUNCTION EVALUATIONS   L = 1591

IFAIL = 0. TERMINATION CRITERION:
NUMBER OF ITERATIONS EQUALS 20

GLOPT TERMINATED

– –
– –

9.13 MIG1: The global method of Monte Carlo (uniform random search)

PURPOSE.   To find the global minimum of a continuous function (9.1.1) of N variables defined on the rectangular parallelepiped (9.1.2).

RESTRICTIONS.   The restrictions are similar to those in all other global methods except that in the case of MIG1 the amount of auxilliary calculation is minimal, so the method can be recommended to minimize very simple functions, say less than 1 sec. of CPU time, when the convergence in probability is sufficient.

ACCURACY.   The method converges in probability to the global minimum of continuous functions. The average deviation is considerably greater compared with global methods with the same number of function evaluations.

HOW TO USE THE METHOD.

CALL MIG1 (X, A, B, N, FM, IPAR, IPA)

where the input is: A, B, N, IPAR, IPA

and the output is X, FM.

In the main program the following arrays should be described:

A(N), B(N), IPAR(30), X(N)
$N \leq 100$
IPAR(1) = IPR is printing parameter
IPAR(2) = M is the number of function evaluations.

EXAMPLE. The program locates the global minimum of the multimodal function (9.5.1).

IPR = 0, M = 10000.

```
     DIMENSION X(2), A(2), B(2),  IPAR(30),
     DATA N,  IPA/2,  0/, A/ – 0.25,  – 0.125/, B/0.5, 0.625/
     DATA IPAR/0,10000, 28*0/
     CALL MIG1 (X, A, B, N, FM, IPAR, IPA)
     STOP
     END
```

```
 FUNCTION FI(X, N)
 DIMENSION X(N)
 FI=FURASN (X, N)
 RETURN
 END
```

MIG1

INITIAL DATA

NUMBER OF VARIABLES   N = 2
PRINTING PARAMETER   IPR = 0
NUMBER OF FUNCTION EVALUATIONS   M = 10000

VECTOR OF LOWER BOUNDS (A) FOR X
$-0.2500000E\ 00$ $-0.12500000E\ 00$
VECTOR OF UPPER BOUNDS (B) FOR X
$0.5000000E\ 00$ $0.62500000E\ 00$

RESULTS

OPTIMAL FUNCTION VALUE FM = $-0.19982195E\ 01$ OBTAINED IN
NR = 1947
OPTIMAL POINT
$0.10454655E-02$ $-0.31354427E-02$

NUMBER OF FUNCTION EVALUATIONS   L = 10000

MIG1 TERMINATED

9.14 MIG2: The modified version of MIG1

This is exactly the same method as MIG1 except that the co-ordinates of the points of all M function evaluations are placed in the array XN of length NM = N * M and, consequently N ≤ 20. The corresponding values of functions are placed into the array /BS1/Y(1000).

HOW TO USE THE METHOD.

CALL MIG2 (X, A, B, N, XN, NM, FM, IPAR, IPA)

where the input is: A, B, N, NM, IPAR, IPA

and the output is X, XN, FM.

In the main program the following arrays should be described:

A(N), B(N), XN(NM), IPAR(30), X(N)
N ≤ 20

IPAR(2) = M is the number of function evaluations, M ≤ 1000.

9.15 EXTR: The global one-dimensional method by Zilinskas

PURPOSE. To locate the global minimum of a continuous function of one variable on a closed interval.

RESTRICTIONS. Restrictions are similar to those of multi-dimensional global methods. The local search is included into EXTR.

ACCURACY. The method provides the minimal average deviation from the minimum under the assumption that the objective function can be regarded as a sample of a Wiener process and converges to the minimum of continuous functions.

HOW TO USE THE METHOD.

CALL EXTR (X, A, B, FM, IPAR, PAR, IPA, IPAA)

where the input is: A, B, IPAR, PAR, IPA, IPAA
and the output is: FM, X

IPAR(1) = IPR is printing parameter
IPAR(2) = M is the maximal number of function evaluations, M ≤ 500 .
IPAR(3) = LT is the number of initial points which are random uniformly distributed
points, LT ≥ 6, recommended LT = 6.
PAR(1) = EPS1 is the accuracy of minimization,
PAR(2) = EPS2 is the accuracy of the point to be minimized.

In the main program the following arrays should be described:
IPAR(30), PAR(30).

EXAMPLE. The program locates the minimum of the multimodal function

$$f(x) = -\sum_{i=1}^{5} i \sin ((i + 1)x + i)$$

with $x \in [-10, 10]$.

IPR = 0, M = 200, LT = 6, EPS1 = 10^{-6}, EPS2 = 10^{-6}.

```
      DIMENSION IPAR(30),  PAR(30)
      DATA IPA, IPAA/0, 0/, A, B/-10., 10./
       DATA IPAR/0, 200, 6, 27*0/, PAR/2*1.E-6, 28*0./
      CALL EXTR (X, A, B, FM, IPAR, PAR, IPA, IPAA)
      STOP
      END

      FUNCTION FI(X, N)
      A = 0.
      DO 2 I = 1,5
      AI = FLOAT(I)
    2 A = A - AI * SIN((AI + 1.) * X + AI)
      FI = A
      RETURN
      END
```

EXTR

INITIAL DATA

PRINTING PARAMETER   IPR = 0
MAXIMUM NUMBER OF FUNCTION EVALUATIONS   M = 200
NUMBER OF FUNCTION EVALUATIONS
                    FOR PARAMETER ESTIMATION   LT = 6
ACCURACY OF OPTIMAL FUNCTION VALUE EPS1 = 0.10000000E–07
ACCURACY OF OPTIMAL POINT   EPS2 = 0.10000000E–07

LOWER BOUND (A) FOR X = – 0.10000000E 02
UPPER BOUND (B) FOR X = 0.10000000E 02

– –
– –

RESULTS

OPTIMAL FUNCTION VALUE   FM = – 0.12031261E 02
OPTIMAL POINT   XM = – 0.67745743E 01
              LOCAL OPTIMA
        POINT   FUNCTION VALUE
– 0.10000000E 02   – 0.26305466E 01
– 0.67745743E 01   – 0.12031261E 02
– 0.17255497E 01   – 0.94947052E 01
– 0.49139261E 00   – 0.12031249E 02
  0.45579252E 01   – 0.94947214E 01
  0.57918062E 01   – 0.12031260E 02

NUMBER OF FUNCTION EVALUATIONS   L = 157

IFAIL = 0. TERMINATION CRITERION:
PROBABILITY OF FINDING THE GLOBAL OPTIMUM WITH GIVEN
ACCURACY IS MORE THAN 0.95

EXTR TERMINATED

– –
– –

9.16   MIVAR4: The local method of variable metrics by Tieshis

PURPOSE. To locate the local minimum of a differentiable function (9.1.1) defined on the rectangular parallelepiped (9.1.2).

RESTRICTIONS. Only the local minimum of differentiable functions can be found. Both numerical and analytical differentiation can be used. In the analytical case the subroutine of differentiation should be provided.

ACCURACY. Arbitrarily close approaches to the minimum can be made depending on parameters EPS.

HOW TO USE THE METHOD.

   CALL MIVAR4 (X, A, B, N, HES, NH, FM, IPAR, PAR, IPA, IPAA)

where the input is: X, A, B, N, NH, IPAR, PAR, IPA, IPAA

and the output is X, FM,

HES is a working array of length NH which contains the elements of inverse Hessian, where $NH = N(N + 1)/2$.

In the main program the following one-dimensional arrays should be described:

A(N), B(N), X(N), HES(NH), IPAR(30), PAR(30).
$N \leq 100$
IPAR(1) = IPR is printing parameter
IPAR(2) = M is the maximal number of function evaluations,
IPAR(3) = NSTOP is the stopping parameter if in the sequence of length NSTOP of iterations the values of function are decreasing less than EPS1 per iteration then the procedure stops. Recommended NSTOP > 1
IPAR(4) = IMAX is the maximal number of iterations,
PAR(1) = XEPS is the step length tolerance,
PAR(2) = EPS is the norm of gradient tolerance,
PAR(3) = EPS1 is the function decreasing tolerance,
PAR(4) = DELT is the length of initial step of numerical differentiation.
 If the partial derivatives are defined analytically then the following suboutine is to be used to define the gradient:

SUBROUTINE GRABP1(X, N, A, B)

```
DIMENSION X(N), A(N), B(N)
COMMON /B3/GR(100)
GR(N) = ∂f/∂xₙ
RETURN
END
```

The values of gradient are put in the one-dimensional array GR of length 100 by the common block /B3/.

EXAMPLE. The program locates the local minimum of the function (9.5.1).

IPR = 0,   M = 100, NSTOP = 2, IMAX = 100, XEPS = 100, EPS = 10^{-4}, EPS1 = 10^{-4}, DELT = 10^{-4}. Co-ordinates of initial point = $(-0.1; 0.1)$

```
DIMENSION X(2), A(2), B(2), HES(3),   IPAR(30),  PAR(30)
DATA IPAR/0, 100, 2, 100, 26*0/, PAR/100.,3*1.E-4, 26*0./,
X/ -0,1., 0.1/
 DATA N, NH, IPA, IPAA/2, 3, 0, 0/,  A/ - 0.25,  - 0.125/, B/0.5, 0.625/
CALL MIVAR4 (X, A, B, N, HES, NH, FM, IPAR, PAR, IPA, IPAA)
STOP
END

FUNCTION FI(X, N)
DIMENSION X(N)
FI=FURASN (X, N)
RETURN
END
```

MIVAR4

INITIAL DATA

NUMBER OF VARIABLES	N = 2
PRINTING PARAMETER	IPR = 0
MAXIMUM NUMBER OF FUNCTION EVALUATIONS	M = 100
NUMBER OF SMALL FUNCTION CHANGE RECURRENCE	NSTOP = 2
MAXIMUM NUMBER OF ITERATIONS	IMAX = 100
SMALL STEP TOLERANCE	XEPS = 0.10000000E 03
GRADIENT NORM TOLERANCE	EPS = 0.10000000E–05
FUNCTION CHANGE TOLERANCE	EPS1 = 0.10000000E–05

DIFFERENTIATION STEP   DELT = 0.10000000E–05
VECTOR OF LOWER BOUNDS (A) FOR X
– 0.2500000E 00   – 0.12500000E 00
VECTOR OF UPPER BOUNDS (B) FOR X
0.5000000E 00   0.62500000E 00

STARTING POINT

– 0.99999999E–01   0.99999999E–01

FUNCTION VALUE F = 0.47440255E 00

– –
– –

RESULTS

OPTIMAL FUNCTION VALUE   FM = – 0.20000000E 01
NORM OF CONSTRAINED GRADIENT   = 0.37390131E–02
OPTIMAL POINT
– 0.48235106E–05   0.17049488E–05

NUMBER OF FUNCTION EVALUATIONS   L = 42
NUMBER OF ITERATIONS   NR = 6

IFAIL = 1. TERMINATION CRITERION: CHANGE OF FUNCTION,
LESS THAN EPS1, OCCURED NSTOP TIMES

MIVAR4 TERMINATED

– –
– –

9.17 REQP: The local method of recursive quadratic programming by Biggs

PURPOSE. To locate the local minimum of a differentiable function (9.1.1) with nonlinear constraints

RESTRICTIONS. Only the local minimum of a differentiable function with nonlinear constraints can be found. Both numerical and analytical differentiation can be used. In the analytical case the subroutine of differentiation should be provided. The user should also provide the subroutine of constraints.

ACCURACY. Arbitrarily close approaches to the minimum can be made depending on parameters EPS.

HOW TO USE THE METHOD.

  CALL REQP (X, H, Q, GC, N, FM, IPAR, PAR, IPA, IPAA)

where the input is: X, N, IPAR, PAR, IPA, IPAA

and the output is X, FM,

The working arrays are: H, Q, GC.

In the main program the following arrays should be described:

X(N), Q(N,N), H(N,N), GC(100, N), IPAR(30), PAR(30).
$N \leq 100$
IPAR(1) = IPR is printing parameter
IPAR(2) = IMAX is the maximal number of iterations,
IPAR(3) = NC is the number of equality constraints,
IPAR(4) = NIC is the number of inequality constraints,
PAR(1) = R1 is the penalty parameter, recommended R1 = 1.
PAR(2) = SCALE is the scale parameter of penalty function,
              recommended $0.1 \leq SCALE \leq 0.75$,
PAR(3) = DELTA is the step length of numerical differentiation,
              recommended $10^{-2} \geq DELTA \geq 10^{-6}$

PAR(4) = EPS is the accuracy parameter, recommended $10^{-2} \geq EPS \geq 10^{-6}$.

 Constraints should be represented by the subroutine

  CONSTR(X, N, G, MC)

where G is a one-dimensional array of length MC which contains the constraints at the point X, MC is the number of constraints.
    The equality constraints should be represented at the beginning of array G. In the case of analytical differentiation the derivatives should be represented by the subroutine

```
      SUBROUTINE CALGRD (X, N, G, MC, GC)

      COMMON /B3/GR(100)
      DIMENSION X(N), GC(100,N), G(MC)
      DO 1 I = 1, N
       GR(I) = ∂f/∂xᵢ
      DO 1 J = 1, MC
    1  GC(J, I) = ∂gⱼ/∂xᵢ
      RETURN
      END
```

The values of gradient are put in the one-dimensional array GR of length 100 by the common block /B3/. The values of gradients of constraints are put in the two-dimensional array GC of dimension 100 * N in accordance with the following formula

$$GC(J, I) = \partial g_j / \partial x_i$$

where g_j is the constraint j.

EXAMPLE. The program locates the local minimum of the function

$$f(x) = 4x_1 - x_2^2 - 12$$

with constraints

$$25 - x_1^2 - x_2^2 = 0$$

$$10x_1 - x_1^2 + 10x_2 - x_2^2 - 34 \geq 0$$

$$x_1 \geq 0$$

$$x_2 \geq 0$$

IPR = 0, IMAX = 50, R1 = 1, SCALE = 0.25, DELTA = 10^{-4}, EPS = 10^{-4}, NC = 1, NIC = 3. The initial point is (1, 1)

```
DIMENSION X(2), H(2, 2), Q(2, 2), GC(100, 2), IPAR(30), PAR(30)
DATA N, IPA, IPAA/2, 0, 0/, X/1., 1./
DATA IPAR/0, 50, 1, 3, 26*0/, PAR/1., 0.25, 2*1.E–4, 26*0./

CALL REQP (X, H, Q, GC, N, FM, IPAR, PAR, IPA, IPAA)
STOP
END

FUNCTION FI(X, N)
DIMENSION X(N)
FI=4.*X(1) – X(2)**2 – 12.
RETURN
END

SUBROUTINE CONSTR (X, N, G, M)
DIMENSION X(N), G(M)
G(1) = 25. – X(1)**2 – X(2)**2
G(2) = 10. *X(1) – X(1)**2 + 10.*X(2) – X(2)**2 – 4.
G(3) = X(1)
G(4) = X(2)
RETURN
END
```

REQP

INITIAL DATA

NUMBER OF VARIABLES	N = 2
PRINTING PARAMETER	IPR = 0
MAXIMUM NUMBER OF ITERATIONS	IMAX = 50
NUMBER OF EQUALITY CONSTRAINTS	NC = 1
NUMBER OF INEQUALITY CONSTRAINTS	NIC = 3
PENALTY PARAMETER	RI = 0.10000000E 01
SCALING PARAMETER	SCALE = 0.25000000E 00
DIFFERENTIATION STEP	DELTA = 0.10000000E–05
TOLERANCE LEVEL	EPS = 0.10000000E–05

STARTING POINT = INFEASIBLE

0.10000000E 01   0.10000000E 01

FUNCTION VALUE F = – 0.90000000E 01
CONSTRAINTS
0.23000000E 02   –0.16000000E 02   0.10000000E 01   0.10000000E 01

– –
– –

RESULTS

OPTIMAL FUNCTION VALUE   FM = – 0.31992371E 01

OPTIMAL POINT
 0.10012760E 01   0.48987226E 01

CONSTRAINTS

– 0.45776367E–04 – 0.45776367E–04   0 .1001277E 01   0.48987236E 01

LAGRANGE MULTIPLIERS

0.10003557E 01   0.74965036E 00   – 0.46047888E 01   0.0

NUMBER OF ITERATIONS	K = 11
NUMBER OF FUNCTION EVALUATIONS	L = 45
NUMBER OF GRADIENT EVALUATIONS	LG = 11
NUMBER OF ACTIVE CONSTRAINTS	MA = 2

NUMBERS OF ACTIVE CONSTRAINTS
     1   2

IFAIL = 0. TERMINATION CRITERION: NORMS OF GRADIENTS
LESS THAN EPS

REQP TERMINATED

– –
– –

9.18   FLEXI: The local simplex method by Nelder and Mead

PURPOSE. To locate the local minimum of a nondifferentiable function (9.1.1) with nonlinear constraints.

RESTRICTIONS. Only the local minimum of a function with constraints can be found. The user should provide the subroutine of constraints.

ACCURACY. Convergence to the minimum is not provided but usually the accuracy satisfies the practical needs if the number of iterations is large enough.

HOW TO USE THE METHOD.

　　　CALL FLEXI (X, N, FM, IPAR, PAR, IPA, IPAA)

where the input is: X, N, IPAR, PAR, IPA, IPAA,

and the output is X, FM.

In the main program the following arrays should be described:

X(N), IPAR(30), PAR(30).
$N \leq 20$
IPAR(1) = IPR is printing parameter,
IPAR(2) = M is the maximal number of function evaluations,
IPAR(3) = NC is the number of equality constraints,
IPAR(4) = NIC is the number of inequality constraints, $NC + NIC \leq 100$.
PAR(1) = DELTA is the dimension of the initial simplex.
　　　　　　　Recommended DELTA = 0.2 min (B(I) – A(I))
PAR(2) = EPS is the stopping accuracy, recommended EPS = 10^{-5} or EPS = 10^{-6}.

Constraints should be represented by the subroutine

　　　CONSTR(X, N, G, MC)

where G is the one-dimensional array of length MC which contains the values of constraints at the point X and MC is the number of constraints.
The equality constraints should be represented at the beginning of array G.

EXAMPLE. The program locates the local minimum of the function

$$f(x) = 4x_1 - x_2^2 - 12$$

with constraints

$$25 - x_1^2 - x_2^2 = 0$$

$$10x_1 - x_1^2 + 10x_2 - x_2^2 - 34 \geq 0$$

$$x_1 \geq 0$$

$$x_2 \geq 0$$

IPR = 0, M = 200, NC = 1, NIC = 3.
DELTA= 0.3, EPS = 10^{-5}.
The initial point is (1, 1)

```
      DIMENSION X(2), IPAR(30), PAR(30)
      DATA N, IPA, IPAA/2, 0, 0/, X/1., 1./
      DATA IPAR/0, 200, 1, 3, 26*0/, PAR/ 0.3, 1.E-5, 28*0./

      CALL FLEXI (X, N, FM, IPAR, PAR, IPA, IPAA)
      STOP
      END

      FUNCTION FI(X, N)
      DIMENSION X(N)
      FI=4.*X(1) - X(2)**2 - 12.
      RETURN
      END

      SUBROUTINE CONSTR (X, N, G, M)
      DIMENSION X(N), G(M)
      G(1) = 25. - X(1)**2 - X(2)**2
      G(2) = 10. *X(1) - X(1)**2 + 10.*X(2) - X(2)**2 - 34.
      G(3) = X(1)
      G(4) = X(2)
      RETURN
      END
```

FLEXI

INITIAL DATA

NUMBER OF VARIABLES	N = 2
PRINTING PARAMETER	IPR = 0
MAXIMUM NUMBER OF FUNCTION EVALUATIONS	M = 200
NUMBER OF EQUALITY CONSTRAINTS	NC = 1
NUMBER OF INEQUALITY CONSTRAINTS	NIC = 3

SIZE OF INITIAL POLYHEDRON   DELTA = 0.30000000E 00
DESIRED CONVERGENCE   EPS   = 0.10000000E–06

STARTING POINT

0.10000000E 01   0.10000000E 01

FUNCTION VALUE F = – 0.90000000E 01
CONSTRAINTS
0.23000000E 02   –0.16000000E 02   0.10000000E 01   0.10000000E 01

STARTING POINT = INFEASIBLE
CALCULATED FEASIBLE STARTING POINT
0.25836763E 01   0.43556452E 01

FUNCTION VALUE F = – 0.206369170E 02
CONSTRAINTS
– 0.64701843E 00 0.97461853E 01   0.2583676E 01   0.43556452E 01

– –
– –

RESULTS

OPTIMAL FUNCTION VALUE   FM = – 0.31992172E 02

OPTIMAL POINT
 0.10013027E 01   0.48987141E 01

CONSTRAINTS

0.0   0.15258789E–03   0 .10013027E 01   0.48987141E 01

NUMBER OF FUNCTION EVALUATIONS   L = 53

NUMBER OF ITERATIONS   NR = 24

IFAIL = 0. TERMINATION CRITERION: TOLERANCE CRITERION
LESS THAN EPS

FLEXI TERMINATED

‒ ‒
‒ ‒

9.19 LBAYES: The local Bayesian method by Mockus

PURPOSE. To locate the minimum of a unimodal function with noise on the rectangular parallelepiped (9.1.2).

RESTRICTIONS. Only the minimum of a unimodal function is found

ACCURACY. Arbitrarily close approaches to the minimum can be made with probability 1 when the number of iterations is large enough. The Bayesian step length provides the minimal average deviation in accordance with a given statistical model, see Mockus (1984[a]).
      The number of iterations should be increased sharply if we wish to make the average error considerably less than the level of noise.

HOW TO USE THE METHOD.

      CALL LBAYES (X, A, B, N, F, IPAR, PAR, IPA, IPAA)

where the input is: A, B, N, IPAR, PAR, IPA, IPAA,

and the output is: X, F, XM, FM

where

X is the last point,
F is the value of the function at point X,

XM is an array of length N which defines the point of minimum of the function,
FM is the minimum
Array XM and FM are defined by the common block COMMON/LAIK/FM,
XM(100).

In the main program the following arrays should be defined:

A(N), B(N), IPAR(30), PAR(30), X(N),
N $\leq$ 100
IPAR(1) = IPR is printing parameter,
IPAR(2) = M is the number of iterations,
IPAR(3) = NIPA is the number of integer variables, they should be at the beginning
of the array X,
PAR(1) = ANIU is the rate of decreasing of the differentation step,
PAR(2) = BETA is the rate of decreasing of the iteration step,
$\qquad$ recommended BETA = 1. – 2*ANIU, ANIU = 0.01 to 0.1

In terms of expression (7.3.31) ANIU = ν, BETA = $1 - \nu - \alpha$, to provide the
convergence $\alpha \geq 0$, $\nu > 0$, $\alpha + \nu < 0.5$, $\nu - \alpha > 0$.

EXAMPLE. The program locates the local minimum of the following function with
noise

$$f(x) = \sum_{i=1}^{n} (x_i^3/6000 + x_i^2/200) + \xi$$

with $n = 2$, $x \in [-10, 10]$.

Here IPR = 0, M = 20, NIP = 1, ANIU = 0.05, BETA = 0.9, and ξ is the random
variable uniformly distributed in the interval $[-0.5, 0.5] * 0.046 \sqrt{(n/2)}$, by the real
function ATS(1) from this package. The initial point is (5, 5).

```
DIMENSION X(2), A(2), B(2), IPAR(30), PAR (30)
DATA N, IPA,  IPAA/2, 0, 0/, A/– 10., – 10./, B/2*10./, X/2*5.)
DATA IPAR/0, 20, 1, 27*0/, PAR/0.05, 0.9, 28*0./
CALL LBAYES (X, A, B, N, F, IPAR, PAR, IPA, IPAA)
STOP
END
```

```
      FUNCTION FI(X, N)
      DIMENSION X(N)
      Y = 0.
      DO 10 I = 1, N
   10 Y = Y+X(I)**3/6.E + 3 + X(I)**2/2.E + 2
      A = ATS(1) - 0.5
      Y = Y + 0.046*A*SQRT(N/2.)
      FI=Y
      RETURN
      END
```

LBAYES

INITIAL DATA

NUMBER OF VARIABLES	N	= 2
PRINTING PARAMETER	IPR	= 0
NUMBER OF ITERATIONS	M	= 20
NUMBER OF INTEGER VARIABLES	NIPA	= 1

RATE OF TRIAL STEP DECREASING   ANIU = 0.50000000E–01
RATE OF ITERATION STEP DECREASING   BETA = 0.90000000E 00

VECTOR OF LOWER BOUNDS (A) FOR X
− 0.1000000E 02   − 0.10000000E 02
VECTOR OF UPPER BOUNDS (B) FOR X
0.1000000E 02   0.10000000E 02

STARTING POINT
0.50000000E 01   0.50000000E 01
FRUNCTION VALUE F = 0.28995705E 00

— —
— —

RESULTS

LAST FUNCTION VALUE F = − 0.38817651E–02
LAST POINT
− 0.95811367E– 01   −.24840391E 00

OPTIMAL FUNCTION VALUE FM = $-0.83632320E-02$ OBTAINED IN NR = 9
OPTIMAL POINT
$-0.10349190E$ 02 $-0.17454523E$ 00
NUMBER OF FUNCTION EVALUATIONS   L = 406

LBAYES TERMINATED

– –
– –

9.20   ANAL1: The method of analysis of structure by Shaltenis

PURPOSE. To define the variables or pairs of variables which have the greatest influence on the accuracy of optimization.

RESTRICTIONS. The assumption is made that the search procedure is nearly uniform. The program was also successfully used for nonuniform optimization procedures. No special assumptions are made about the function except that it is continuous and satisfies some general conditions concerning average behaviour.

ACCURACY. Arbitrarily close estimation of the structural characteristics can be made depending on the results of function evaluation.

HOW TO USE THE METHOD.

CALL ANAL1 (A, B, N, XN, XM, NM, IPAR, IPA)

where the input is: A, B, N, XN, NM, IPAR, IPA.

The working array is XM

In the main program the following arrays should be described:

A(N),B(N), XN(NM), XM(NM), IPAR(30)
$N \leq 20$
IPAR(1) = IPR is printing parameter,
IPAR(2) = M is the number of function evaluations, $10 \leq M \leq 300$
IPAR(3) = NH is the number of harmonics, $NH \leq 7$, recommended NH = 7,
IPAR(4) = NSF is the maximal number of variables to be defined, $NSF \leq 30$. NSF depends on M. Larger M permits larger NSF.

IPAR(5) = INP is the parameter of structural analysis:

> if INP = 1 then only single variables are considered,
>
> if INP = 2 then pairs of variables are also included into the analysis of importance

The routine ANAL1 can be used if the number $L (L \geq M)$ of evaluations of the function are performed and the following arrays are defined:

/BSI/Y(1000) are values of the function,

XN(NM) are the co-ordinates of corresponding points of evaluation.

To define the arrays the routine MIG2 can be used.

EXAMPLE. The program performs the analysis of structure of the function (9.5.1).

IPR = 0, M = 200, NH = 7, NSF = 8, INP = 2.

```
DIMENSION X(2), A(2), B(2), XN(400), XM(400), IPAR(30)
DATA N, NM, IPA/2, 400, 0/, A/ – 0.25, – 0.125/, B/0.5, 0.625/
DATA IPAR/–1, 200, 0, 200, 7, 8, 2, 23*0/
CALL MIG2 (X, A, B, N, XN, NM, FM, IPAR, IPA)
IPA = 2
CALL ANAL1 (A, B, N, XN, XM, NM, IPAR, IPA)
STOP
END

FUNCTION FI(X, N)
DIMENSION X(N)
 FI=FURASN (X, N)
RETURN
END
```

ANAL1

INITIAL DATA

NUMBER OF VARIABLES	N	= 2
PRINTING PARAMETER	IPR	= 0
NUMBER OF FUNCTION EVALUATIONS	M	= 200
NUMBER OF HARMONICS	NH	= 7
NUMBER OF SELECTED FACTORS	NSF	= 8
INTERACTION PARAMETER	INP	= 2

VECTOR OF LOWER BOUNDS (A) FOR X
$-0.2500000E\ 00$ $-0.12500000E\ 00$
VECTOR OF UPPER BOUNDS (B) FOR X
$0.5000000E\ 00$ $0.62500000E\ 00$

_ _
_ _

RESULTS
VARIABLES OR PAIRS OF VARIABLES   DEGREE OF INFLUENCE
              X2 $0.40290910E\ 00$
              X1 $0.32305169E\ 00$

ANAL1 TERMINATED

_ _
_ _

9.21 Portability routines

I1MACH, R1MACH defines the machine constants. Those routines correspond to the PORT subroutine library. They are described in the BELL laboratories computing science technical report 47 by P.A. Fox, A.D. Hall and N.L Schryer.

INTEGER FUNCTION I1MACH(I):

	I/O unit numbers	Machine constants for the VAX-II with FORTRAN IV-PLUS
I1MACH(1)	is the standard input unit,	5
I1MACH(2)	is the standard output unit,	6
I1MACH(3)	is the standard punch unit,	7
I1MACH(4)	is the standard error message unit,	6
	Words	
I1MACH(5)	is the number of bits per integer storage unit	32
I1MACH(6)	is the number of characters per integer storage unit,	4
	Integers	
I1MACH(7) = A	is the base,	2
I1MACH(8) = S	is the number of the base A digits,	31
I1MACH(9) = A**S−1	is the largest magnitude,	2147483647
	Floating point numbers	
I1MACH(10) = B	is the base,	2
	Single precision	
I1MACH(11) = T	is the number of the base B digits,	24
I1MACH(12) = EMIN	is the smallest exponent E,	− 127
I1MACH(13) = EMAX	is the largest exponent E,	127
	Double precision	
I1MACH(14) = T	is the number of the base B digits,	56
I1MACH(15) = EMIN	is the smallest exponent E,	− 127
I1MACH(16) = EMAX	is the largest exponent E,	127

REAL FUNCTIONS R1MACH(I):

Single precision machine constants

R1MACH(1) = B**(EMIN – 1)   is the smallest positive magnitude   Z00000080
R1MACH(2)
  = B**EMAX*(1 – B**(–T))   is the largest magnitude   ZFFFF7FFF
R1MACH(3) = B**(– T)   is the smallest relative spacing   Z00003480
R1MACH(4) = B**(1 – T)   is the largest relative spacing   Z00003500
R1MACH(5) = LOG10(B)   Z209B3F9A

Double precision machine constants

D1MACH(1) = B**(EMIN – 1)   is the smallest positive magnitude   Z00000080
                                                                    Z00000000
D1MACH(2)   ZFFFF7FFF
  = B**EMAX*(1 - B**(– T))   is the largest magnitude   ZFFFFFFFF
D1MACH(3) = B**(– T)   is the smallest relative spacing   Z00002480
                                                                    Z00000000
D1MACH(4) = B**(1– T)   is the largest relative spacing   Z00002500
                                                                    Z00000000
D1MACH(5) = LOG10(B)   Z209A3F9A
                                                                    ZCFFA84FB

The list of machine constants should be fixed by operator DATA in the same order as shown here.

References

Aczel, J. (1966) *Lectures on Functional Equations and their Applications*, Academic Press, New York.

Anderson, T.W. (1958) *An Introduction to Multivariate Statistical Analysis*, 2nd. ed., Wiley, New York.

Archetti, F. (1975) 'A sampling technique for global optimization'. In: *Towards Global Optimization 1*, ed. Dixon, L.C.W. and Szego, G.P., North-Holland, 158-165.

Barauskas, A., Zilinskas, A., Piliavskij, V., Shulman, V., Jushkene, E. (1980) 'Investigation of the optimization problems of the synthesis of pigmental composisitions'. In: *Optimal Decision Theory*, Inst. of Math. and Cybernetics, Vilnius, 6, 57-74 (in Russian).

Barauskas, A. (1984) 'Linear constraints in the problem of multiextremal optimization'. In: *Optimal Decision Theory*, Inst. of Math. and Cybernetics, Vilnius, 10, 22-33 (in Russian).

Barauskas, A. (1984) 'Package of applied software for OPTIMUM to solve multimodal problems of design'. Information leaflet on scientific-technical achievement, No. 84-60 Series 50, Lith. MTII. (in Russian).

Barbee, H.C.P., Boender, C.G.E., Rinnoy Kan, A.H.G., Smith, R.I., Telgen, J. (in print) 'Hit-and-run Algorithm for the indication and identification of nonredundant linear constraint inequalities'.

Bashkis, A., Zanevicius, D., Kostetskij, Ch., (1981) *The analysis of integrated circuits in nonisothermic conditions using the transconde models of transistors*. Preprint No. 10. Inst. of semiconductor physics of the Academy of Sciences of the Lithuanian S.S.R., Vilnius (in Russian).

Bashkis, A., Zanevicius, D., Mockus, J.B., Valevichene, J.P., Dailydenas, V.I. (1982) 'The calculation of the relation of the yield of integrated circuits on the widths of diffused resistors'. Proc. of High Schools of Lithuanian S.S.R., *Radioelectronics*, 18, 98 (in Russian).

Bashkis, A. (1984) 'The optimization of the design of integrated circuits to increase the yield'. In :*Mathematical and computer simulation in microelectronics*, Vilnius, 17-24 (in Russian).

Betro, B., De Biase, L. (1976) 'A recursive spline technique for uniform approximation of sampled data'. *Quaderni del Dipartamento di Ricerca Operativa e Scienze Statistiche*, **A31**, Universita di Pisa.

Betro, B. (1981) 'Bayesian testing of nonparametric hypotheses and its application to global optimization'. Technical Report, CNR-IAMI, Italy.

De Biase, L., Frontini, F. (1978) 'A stochastic method for global optimization: its structure and numerical performance'. In: *Towards Global Optimization 2*, ed. Dixon, L.C.W. and Szego, G.P., North-Holland, 85-102.

Biggs, M.C. (1974) 'Constrained minimization using recursive quadratic programming: some alternative subproblem formulations'. Numerical Optimization Center. Technical Report No. 51.

Biggs, M.C. (1975) 'Constrained minimization using recursive quadratic programming: some alternative subproblem formulations'. In: *Towards Global Optimization 1*, ed. Dixon, L.C.W. and Szego, G.P., North-Holland, 341-349.

Blum, J.R. (1954) 'Multi-dimensional stochastic approximation method', *Ann. Math. Stat.* **25**, 737-744.

Boender, C.G.E., Rinnooy Kan (1983) 'A Bayesian analysis of the number of cells of a multinomial distribution', *The Statistician* **32**, 240-248.

Chichinadze, V.K. (1967) 'Random search to determine the extremum of functions of several variables', *Engineering cybernetics (Technicheskaya kibernetika)*, No. 1 , 115-123 (in Russian).

Collatz, L. (1964) *Funktionanalysis und numerische mathematik*, Springer-Verlag, Berlin .

De Groot, M. (1970) *Optimal Statistical Decisions*, McGraw-Hill, New York.

Didzgalvis, R., Zilinskas, A., Ragulskis, K., Tieshis, V. (1975) 'Optimal synthesis of a linear shock absorber'. In: *Optimal Decision Theory*, Inst. of Math. and Cybernetics, Vilnius, **4**, 13-26 (in Russian).

Didzgalvis, R., Zilinskas, A., Ragulskis, K., Tieshis, V. (1976) 'On optimal energy transmission through skew impact'. In: *Optimal Descision Theory*, Inst. of Math. and Cybernetics, Vilnius, **2**, 9-27 (in Russian).

Dixon, L.C.W. and Szego, G.P. (1978) 'Global optimization: an introduction'. In: *Towards Global Optimization 2*, ed. Dixon, L.C.W. and Szego, G.P., North-Holland, 1-15.

Dixon, L.C.W. and Szego, G.P. (1978) ed. *Towards Global Optimization 2*, North-Holland.

Dvoretzky, A. (1956) 'On stochastic approximation', *Proc. Third Berkeley Symp. on Math. Stat. and Prob.*, **1**, 39-55.

Dzemyda, G. (1981) 'Portability of program libraries'. In: *Optimal Descision Theory*, Inst. of Math. and Cybernetics, Vilnius, **7**, 41-48 (in Russian).

Dzemyda, G. (1982) 'On an extremal grouping'. In: *Optimal Decision Theory*, Inst. of Math. and Cybernetics, Vilnius, **8**, 46-54 (in Russian).

Dzemyda, G. (1983) 'LP-search, taking into account the structure of an extremal problem'. In: *Optimal Decision Theory*, Inst. of Math. and Cybernetics, Vilnius, **9**, 39-44 (in Russian).

Dzemyda, G., Vaitekunas, G., Vyshniauskas, J., Juzefovich, D., Kucherenko, V., Kuziakin, O., Filatov, N. (1984) 'Solution of the problems of optimal design and selection of model parameter values using the package of applied programs MINIMUM'. In: *Optimal Decision Theory*, Inst. of Math. and Cybernetics, Vilnius, **10**, 77-98 (in Russian).

Ermolyev, Yu.M. (1976) *Methods of Stochastic Programming*, Nauka, Moscow (in Russian).

Everitt, B.S. (1978) *Graphical techniques for Multivariate Data*, Heinemann Educational Books, London.

Fine, T.L. (1973) *Theories of Probability*, Academic Press, New York.

Gantmacher, F.R. (1967) *Theory of Matrices*, Nauka, Moscow (in Russian).

Gemulka, J.A. (1978) 'Users experience with Törn's clustering algorithm'. In: *Towards Global Optimization 2*, ed. Dixon, L.C.W. and Szego, G.P., North-Holland, 63-70.

Gemulka, J.A. (1978) 'Two implementations of Branin's method: numerical experience'. In: *Towards Global Optimization 2*, ed. Dixon, L.C.W. and Szego, G.P., North-Holland, 151-164.

Gnedenko, B.V. (1943) 'Sur la distribution limite du terme maximum d'une serie d'eatore, *Ann. Math.* **44**, Nr. 2, 423-453.

Gnedenko, B.V. (1965) *Course of Theory of Probability*, Nauka, Moscow (in Russian).

Grigas, F.J., Gudanavichute, V.V. (1980) 'The calculation of distribution of ampere-turns of magnetic deflection systems'. Proc. of High Schools of Lithuanian S.S.R., *Radioelectronics*, **16**, No. 2, 81 (in Russian).

Grishagin, V.A. (1978) 'Operative charateristics of some algorithms of global search'. In: *Problems of Random Search*, 7,198-206 ed. Rastrigin, L.A., Zinatne, Riga, (in Russian).

Herstein, I.N. and Milnor, J. (1953) 'An axiomatic approach to measurable utility', *Econometrica*, **21**, 291-297.

Himmelblau, D.M. (1972) *Applied nonlinear programming*, McGraw-Hill, New York.

Hochshtein, A. Ya. (1976) *The surface tension of rigid bodies and adsorption*, Nauka, Moscow, (in Russian).

Joshida, T., Osaka, T. (1978) 'The investigation of adsorbtion of hydrogen on a platinum electrode using dynamic measurement of impedance', *Electrochemistry*, **14**, 692-694 (in Russian).

Juzefovich, D., Naumov, Yu., Bereshnov, N., Radvilavichute, J., Shaltenis V., Kuchorenko, V. (1984) 'The investigation of conditions for the regeneration of ferro-copper chloride solutions for etching using the essential variable selection algorithm'. In: *Optimal Decision Theory*, Inst. of Math. and Cybernetics, Vilnius, **10**, 144-154 (in Russian).

Katkauskaite, A.J. (1972) 'Random Fields with independent increments', *Lithuanian Math. Collection* **XII**, No. 4, 75-85 (in Russian).

Katkauskaite, A., Zilinskas, A., (1977) 'On the construction of statistical models of functions under uncertainty'. In: *Optimal Decision Theory*, Inst. of Math. and Cybernetics, Vilnius, **3**, 19-29 (in Russian).

Kramer, H. (1946) *Mathematical Methods of Statistics*, Princeton University Press.

Kuratovski, K. (1966) *Topology 1*, Academic Press, New York, London and Pan.Wyd.Nauk., Warsaw .

Kuratovski, K. (1968) *Topology 2*, Academic Press, New York, London and Pan.Wyd.Nauk., Warsaw.

Kushner, H. (1964) 'A new method of locating the maximum point of an arbitrary multipeak curve in the presence of noise', Trans. ASME, series C, **86**, 97-105.

Loeve, M. (1960) *Probability Theory*, D. Van Nostrand Co. Inc. Princeton, New York.

Marchuk, G.I. (1980) *Mathematical Models in Immunology*, Nauka, Moscow (in Russian).

Mashkovtsev, V.M., Tsibizov, K.N., Jemelin, V.F. (1966) *Theory of Waveguides*, Nauka, Moscow (in Russian).

MINIMUM (1986) The software system for the interactive optimization of multimodal problems 'Algorithms and Programs', No. 6, Register No. 50860000112, GOSFAP, Smolnaia St., 14, Moscow (in Russian).

Mockus, J.B. (1963) 'On an application of the Monte Carlo method in multiextremal and combinatorial problems', Conference Proc., Lectures, Vol IV, *General Problems of the Application of Probabilistic and Statistical Methods*, State Publ. House of Techn. Lit, Ukrainian S.S.R, Kiev, 30-41 (in Russian).

Mockus, J.B. (1964) 'On a method of distribution of trial points in multimodal problems', *Journal vychislitelnoj matematiki i matematicheskoj fiziki* **4**, No. 2, 380-385 (in Russian).

Mockus, J.B. (1965) 'On a sequential procedure for statistical decisions in the extremal problems'. In: *Automat. i Vycisl. Techn.* **10**, Riga, 'Zinatne', 78-101 (in Russian).

Mockus, J.B. (1967) *Multiextremal Problems in Design*, Nauka, Moscow (in Russian).

Mockus, J.B. (1969) 'On a problem of optimal search for an extremum, *Abst. IVth Symposium of Extremal Problems*, Press of Physical-Technical problems of Energetics, Kaunas, p. 5 (in Russian).

Mockus, J.B. (1972) 'On Bayesian methods of search for an extremum', *Automat. i Vycisl. Techn.* **3**, 53-62 (in Russian).

Mockus, J.B. (1977) 'On Bayesian methods for seeking the extremum and their applications'. In: *Information Processing 77*, ed. Gilchrist, B., North-Holland, 195-200.

Mockus, J.B., Tieshis, V., Zilinskas, A. (1978) 'The application of Bayesian methods for seeking the extremum'. In: *Towards Global Optimization 2*, ed. Dixon, L.C.W. and Szego, G.P., North-Holland, 117-130.

Mockus, J.B. (1978) 'The sufficient conditions for convergence of Bayesian methods to a global minimum for any continuous function'. In: *Optimal Decision Theory*, Inst. of Math. and Cybernetics, Vilnius, **4** , 67-89 (in Russian).

Mockus, J., Senkiene, E. (1979) 'On the estimate of the random Gaussian field mean from the observation of dependant random points'. In: *Optimal Decision Theory*, Inst. of Math. and Cybernetics, Vilnius, **4** , 27-38 (in Russian).

Mockus, J. (1980) 'Sufficient conditions for the convergence of the one-dimensional Bayesian method to the global minimum for any continuous function'. In: *Optimal Decision Theory*, Inst. of Math. and Cybernetics, Vilnius, **6**, 9-17 (in Russian).

Mockus, J. (1983) 'The Bayesian approach to global optimization', *2nd Int. Meeting on Bayesian Statistics*, Valencia, Spain, 6-10 September.

Mockus, J.B. (1984) 'On Bayesian approach to stochastic programming', *Abst. Int. Conf. on Stochastic Optimization*, Kiev, 9-16 September, 165-167.

Mockus, J.B. (1984[b]) 'On Bayesian methods in nondifferential and stochastic programming', *Abst. IIASA Workshop on Nondifferential Optimization*, Sopron, Hungary, 17-22 September, 112-117.

Mockus, J. B. (1984[c]) 'The Bayesian approach to global optimization', *Proc. Indian Stat. Inst. Golden Jubilee Conf. on Statistics: Applications of New Directions*, Calcutta 16 Dec. 1981, 405-430.

Mockus, J. B. (1984[d]) 'The Bayesian approach to global optimization', Preprint No. 176, Free University, Berlin, May .

Mockus, J. B. (1984[e]) 'The Bayesian approach to global optimization', Preprint No. 175, Free University, Berlin, May.

Mockus, J. B., Zukauskaite, L., Lideikis, T. (1987) 'The Bayesian approach to stochastic approximation procedures and their applications'. In: *Optimal Decision Theory*, Inst. of Math. and Cybernetics, Vilnius, **2**, 71-77 (in Russian).

Mockus, J. B., Mockus, L. B. (1987) 'Some algorithms of global and multiobjective optimization and their computer implementation'. In: *Optimal Decision Theory*, Inst. of Math. and Cybernetics, Vilnius, **2**, 54-70 (in Russian).

Neveu, J. (1964) *Bases mathématique de calcul des probabilités*, Masson et Cie, Paris.

Nikolajev P., Pochnikaev G., Tieshis, V. (1984) 'Modelling and optimization of waveguide to microstrip line directional couplers'. In: *Optimal Decision Theory*, Inst. of Math. and Cybernetics, Vilnius, **10**, 117-126 (in Russian).

Praporov, A.M., Naumov, Yu.I., Juzefovich, D.K., Kucherenko, V.I., Flerov, V.N. (1979) 'The calculations of the productive cost of the etching of printed circuits with recycling and regeneration'. In: *The Exchange of Experience in the Radio Industry*, **10**, 61-63 (in Russian).

Price, L.W. (1978) 'A controlled random search procedure for global optimization'. In: *Towards Global Optimization* 2, ed. Dixon, L.C.W. and Szego, G.P., North-Holland, 71-84.

Ragulskiene, V.L. (1974) *Vibroshock Systems*, Mintis, Vilnius (in Russian).

Rao, J.S., Joshi, K.K., Das, B.N. (1981) 'Analysis of small aperture coupling between a rectangular waveguide and a microstrip line', *IEEE Trans. on Microwave Theory and Techn.*, **MTT-29**, No. 2, p. 150.

Rastrigin, L.A. (1968) *The stochastic methods of search*, Nauka, Moscow (in Russian).

Savage, L.I. (1954) *Foundations of Statistics*, Wiley, New York.

Senkiene E. (1980) 'Properties of the conditional means and the variance of the Wiener process in the presence of noise and the convergence of the Bayesian optimization algorithms'. In: *Optimal Decision Theory*, Inst. of Math. and Cybernetics, Vilnius, **6**, 18-40 (in Russian).

Shaltenis, V.R., Mockus, J.B. (1963) 'The choice of the optimal development of distribution networks defining the reasonable zone'. In: *Conference Proc., Lectures, Problems of the applications of Probabilistic and Statistical Methods*, State Publ. House of Techn. Lit. Ukrainian S.S.R., Kiev, 80-85 (in Russian).

Shaltenis, V., Varnaite, A. (1975) 'On the method of reducing dimensionality in multiextremal problems'. In: *Theory of Optimal Design*, Inst. of Math. and Cybernetics, Vilnius, **1**, 23-42 (in Russian).

Shaltenis, V., Varnaite, A. (1976) 'Structure of multiextremal problems'. In: *Theory of Optimal Design*, Inst. of Math. and Cybernetics, Vilnius, **2**, 67-78 (in Russian).

Shaltenis, V. (1976) 'The investigation of the efficiency of the LP-search in some multimodal problems'. In: *Optimal Decision Theory*, Inst. of Math. and Cybernetics, Vilnius, **2**, 59-66 (in Russian).

Shaltenis, V.R., Radvilavichute, I. (1977) 'On the separation of main variables in extremal problems'. In: *Optimal Decision Theory*, Inst. of Math. and Cybernetics, Vilnius, **3**, 57-71 (in Russian).

Shaltenis, V.R. (1980) 'The analysis of problems in interactive systems of optimization'. *Proc. Conf. on the Application of Random Search Methods in C.A.D.* Tallin, Valchus, 118-123 (in Russian).

Shaltenis, V.R., Radvilavichute, I. (1980) 'Investigation of the algorithm for the separation of main variables'. In *Optimal Decision Theory*, Inst. of Math. and Cybernetics, Vilnius, **6**, 41-48 (in Russian).

Shaltenis, V.R., Dzemyda, G. (1982) 'The structure analysis of extremal problems using some approximation of characteristics'. In: *Optimal Decision Theory*, Inst. of Math. and Cybernetics, Vilnius, **8**, 124-140 (in Russian).

Shaltenis, V.R., (1982) 'The efficiency of LP-search and the structure of optimization problems'. In: *Optimal Decision Theory*, Inst. of Math. and Cybernetics, Vilnius, **8**, 115-123 (in Russian).

Shepard, D. (1965) 'A two-dimensional interpolation function for irregularly-spaced data'. In: *Proc. 23rd National Conf. ACM*, New York, .517-524.

Sobolj, I.M. (1968) *Multi-dimensional numerical quadrature formulae and Haar functions*, Nauka, Moscow (in Russian).

Sobolj, I.M., Statnikov, R.B. (1981) *The choice of optimal parameters in problems with many objective functions.* Nauka, Moscow (in Russian).

Sobolj, I.M. (1985) 'Points of uniform filling of the multi-dimensional cube'. In : *Mathematics and Cybernetics*, **2**, Znanie Moscow, 1-32 (in Russian).

Strongin, R.G. (1978) *Numerical Methods in Multiextremal Problems.* Nauka, Moscow (in Russian).

Tieshis, V.A. (1975) 'The method of variable metrics for local optimization of functions of many variables with rectangular constraints'. In: *Proc. Conf. on Computers*, Kaunas, 111-114 (in Russian).

Tieshis, V.A. (1979) 'The structure of the package for nonlinear programming'. In *Optimal Decision Theory*, Inst. of Math. and Cybernetics, Vilnius, **5**, 95-102 (in Russian).

Törn, A.A. (1978) 'A search-clustering approach to global optimisation'. In : *Towards Global Optimization 2*, ed. Dixon, L.C.W. and Szego, G.P., North-Holland, 42-69.

Urjasjev, S.P. (1986) 'On adaptive parameter control in stochastic gradient algorithms'. In: *Preprints of 2nd. IFAC Symposium on stochastic control*. Vilnius, Part 11, Moscow, 83-87 (in Russian).

Wasan, M.T. (1969) *Stochastic Approximation*, Cambridge University Press.

Zanevichus, D.J. (1984) 'The nonlinear simulation of integrated circuits'. In: *Mathematical and computer simulation in microelectronics*, Vilnius, 3-16 (in Russian).

Zigliavski, A. (1985) *Mathematical Theory of Global Random Search*, Leningrad University Publishing House, Leningrad (in Russian).

Zilinskas, A. (1978[a]) 'Investigation of multi-dimensional extrapolation under uncertainty'. In: *Optimal Decision Theory*, Inst. of Math. and Cybernetics, Vilnius, **4**, 27-44 (in Russian).

Zilinskas, A. (1978[b]) 'On one-dimensional multimodal minimization', *Trans. 8-th Prague Conf. on Inform. Theory, Stat. Dec. Funct., Proc.* **B**, Academia, Prague, 397-488.

Zilinskas, A. (1978[c]) 'On statistical models for multimodal optimiztion', *Mat. Operationsforsch. Statist., Ser. Statistics*, **9**, No. 2, 255-266.

Zilinskas, A. (1979) 'An axiomatic approach to extrapolation under uncertainty', *Automatika i Telemechanika*, **12**, 66-70 (in Russian).

Zilinskas, A. (1980) 'The use of statistical models for construction of multimodal optimization algorithms'. In: *3rd Czechoslovak-Soviet-Hungarian Seminar on Information Theory*, Czechoslovak Academy of Sciences, Prague, 219-224.

Zilinskas, A. (1981) 'The software for the development of specifications for the central sample of a colour picture tube'. In: *Interactive Technology in CAD*, Tallin, the Kalinin Institute, p. 69.

Zilinskas, A. (1983) 'Axiomatic approach to statistical models and their use in multimodal optimization theory, *Mathematical Programming* **22**, 104-116.

Zilinskas, A. (1985) 'Axiomatic characteristics of a global optimization algorithm and investigation of its search strategy', *Operations Research Letters* **4**, No, 1, 35-39.

Zilinskas, A. (1986) *Global Optimization.*, "Mokslas", Vilnius (in Russian).

Zuev, G.M. (1984) *Mathematical models of diseases and the analysis of experimental data*, Dept. of Computational Mathematics, U.S.S.R. Academy of Sciences, Moscow (in Russian).

THE SOFTWARE FOR GLOBAL OPTIMIZATION FOR IBM PC/XT/AT AND COMPATIBLES

Since the software for global optimization is in portable FORTRAN it can be used in personal computers with corresponding operating systems. However, FORTRAN is not the best language for personal computing where interactive procedures are essential. So another version of the software was designed for the IBM compatible personal computers.

The PC version of the global optimization software was implemented using the programming language 'C', see Mockus (1987). The PC version implements most of the methods of global optimization described in this book, namely BAYES1, UNT, LPMIN, MIG1, MIG2, EXTR, LBAYES, ANAL1. In addition, it implements the methods of global optimization with nonlinear constraints and the method for the multi-objective global optimization, see Mockus (1987). Naturally, in the case of personal computing, the interactive possibilities are provided but users who do not wish to influence the process of optimization interactively can use it completely automatically.

The PC version of the software is a multi-level interactive system. On each level the corresponding 'menu' is presented to the user, who can see and change the parameters of method and input-output data, interactively.

The output can be graphical. At any moment the user can see the process of optimization, and can watch the moving point x_{n+1} on the central two-dimensional projection of the objective function $f(x)$, see Figure A.1.

The FORTRAN version is provided on the disc (inside back cover).

To obtain the PC and updated versions, you should get in touch directly with the author:

Prof. J. B. Mockus,
Institute of Mathematics and Cybernetics,
Academy of Sciences of the Lithuanian SSR,
232600 VILNIUS,
USSR.

GLOBAL
MINIMUM

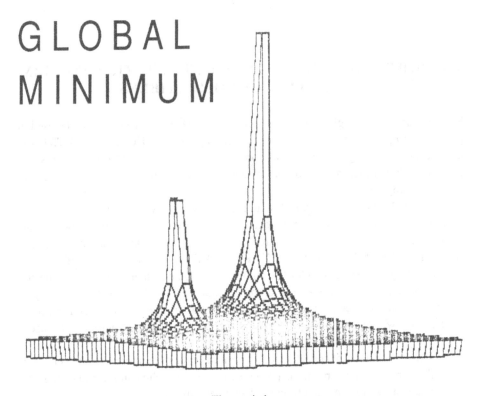

Figure A.1

The graphical output: the central projection of the objective function $f(x)$ as a function
of two variables with fixed values of remaining variables.

One of the first examples of global optimization - the common electric meter,
designed in 1962, see section 8.2 - is still competing successfully in the world market
without major changes in the original configuration.

HOW THE GLOBAL OPTIMIZATION SOFTWARE CAN IMPROVE THE PERFORMANCE OF YOUR CAD SYSTEM

All well known CAD packages have GRAPHICS
Most of them have DATA BASES
Some of them have LINEAR or NONLINEAR PROGRAMMING Procedures
Very few have good software for GLOBAL OPTIMIZATION

WHAT CAN GLOBAL OPTIMIZATION DO FOR YOU?

When you are designing a complicated system you are confronted with the problem of how to choose the values of tens or even hundreds of parameters in order to optimize the system. Good interactive graphics can help a great deal. The usual linear and nonlinear programming software can also be useful if your system can be represented by linear or unimodal functions but this is seldom the case.

So what is the result?

The usual result is that you get a design which looks reasonably good, but it can happen that it is a long way from the best possible decision both in terms of money and in the quality of the product that you are designing.

If your CAD has a properly tuned Global Optimization Package, you can get the design which is really GLOBALLY OPTIMAL. This means that it cannot be significantly improved in terms of the objective which you wish to optimize. It can save you millions of dollars and will give you the edge over your competitors.

The global optimization procedures can work automatically or interactively depending on your problems and preferences.

HOW TO INCLUDE THE GLOBAL OPTIMIZATION INTO YOUR CAD SOFTWARE

There are two basic versions of the global optimization software.

The first is in portable FORTRAN which means that it can run on any computer with standard FORTRAN. It was extensively tested on PDP, VAX and IBM mainframe computers and on personal computers. The second version is in C and is adjusted to IBM PC/XT/AT compatible portable computers.

There are several ways to include the Global optimization software into your CAD software.

The best way naturally is by co-operation during the development of the CAD software. The other way is to include the Global Optimization package as an addition to the existing CAD software.

In both cases the authors of Global Optimization are ready to help you promptly in every possible way.

MACHINE DEPENDENT CONSTANTS OF PORTABLE FORTRAN

In the package all the dependent constants are fixed because all of them could be useful during extensions in the future. For this version of the software no double-precision constants are used so these are omitted.

The constants are given in Table 3, where columns represent different computers and rows correspond to different single-precision constants. The longer constants are written in two rows. The integer N which is repeated L times is denoted as N(Lt).

No.	Constant	VAX-11 FORTRAN IV-PLUS	BURROUGHS 1700	BURROUGHS 5700/6700/7700	CD 6000/7000	CRAY-1	HONEY-WELL 600/6000	IBM 360/370	UNIVAC 1100
1	I1MACH(1)	5	7	5	5	100	5	5	5
2	I1MACH(2)	6	2	6	6	101	6	6	6
3	I1MACH(3)	7	2	7	7	102	43	7	7
4	I1MACH(4)	6	2	6	6	101	6	6	6
5	I1MACH(5)	32	36	48	60	64	36	32	36
6	I1MACH(6)	4	4	6	10	8	6	4	6
7	I1MACH(7)	2	2	2	2	2	2	2	2
8	I1MACH(8)	31	33	39	48	63	35	31	35
9	I1MACH(9)	214748 3647	Z1FFFF FFFFF	0000 7(13t)	0000 7(16t)B	7(21t)B	037(11t)	Z7F(7t)	037(11t)
10	I1MACH(10)	2	2	8	2	2	2	16	2
11	I1MACH(11)	24	24	13	48	47	27	6	27
12	I1MACH(12)	-127	-256	-50	-974	-8192	-127	-6	-128
13	I1MACH(13)	127	255	76	1070	8190	127	63	127
14	R1MACH(1)	Z0000 0080	Z4008 00000	0177 0(12t)	00014 0(15t)B	200004 0(15t)B	04024 0(8t)	Z001 0000	00004 0(8t)
15	R1MACH(2)	ZFFFF 7FFF	Z5FFF FFFFF	007(15t)	3776 7(16t)B	57776 7(15t)6B	0376 7(9t)	Z7FF FFFFF	037(11t)
16	R1MACH(3)	Z0000 3480	Z4E98 00000	01311 0(12t)	16404 0(15t)B	377224 0(15t)B	07144 0(8t)	Z3B 00000	01464 0(8t)
17	R1MACH(4)	Z0000 3500	Z4EA8 00000	01301 0(12t)	16414 0(15t)B	377234 0(15t)B	07164 0(8t)	Z3C1 00000	01474 0(8t)
18	R1MACH(5)	Z209B 3F9A	Z500E 730E8	01157163 034761675	171646420 23241175720B	377774642 3241175720B	0207764642 02324	Z411 34413	0177464 202324

Table A.3

Machine dependent constants of portable FORTRAN

Printed in the United States
By Bookmasters